Volume 1 - Sustainable Meliponiculture with Vernacular Architecture

Author: Abu Hassan Jalil

Jointly published by: The International Bee Research Association,
a Company Limited by Guarantee, 1, Agincourt Street, Monmouth, NP25 3DZ (UK) &
Northern Bee Books, Scout Bottom Farm, Mytholmroyd, Hebden Bridge HX7 SJS (UK).

Obtainable from:
www.ibra.org.uk & www.northernbeebooks.co.uk

© 2023 The International Bee Research Association.

All rights reserved. No part of this publication may be reproduced stored or transmitted in any form or by any means, electronically, mechanically, by photocopying, recording, scanning or otherwise, without the prior permission of the copyright owners.

ISBN: 978-1-913811-15-0

Artwork by DM Design & Print
IBRA Proof Editor - Stuart A. Roberts

Volume 1 - Sustainable Meliponiculture with Vernacular Architecture

Author: Abu Hassan Jalil

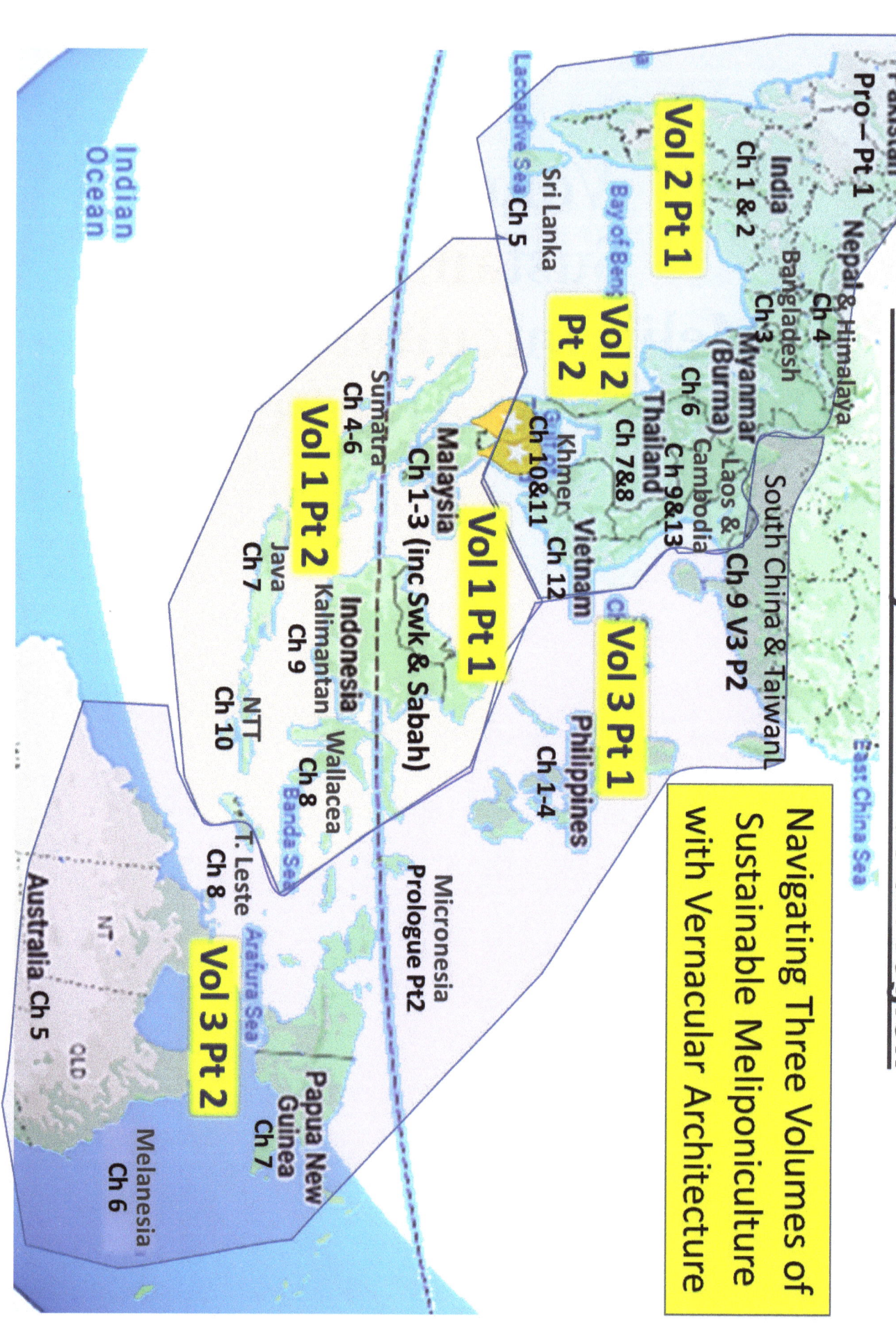

Sustainable Meliponiculture with Vernacular Architecture

Volume 1 - The Malay Archipelago

Synopsis: The current situation in Meliponiculture in the Indo-Malayan Ecozone and Stingless bee housing is haphazard, with no standards in Meliponary and bee housing design. The bee comfort is questionable as beekeepers strive for quick harvests and instant profits from the beekeeping products. Thermal comfort is a term of grave concern regarding conservation in Meliponiculture. Often enough, enthusiastic new beekeepers are not overly worried about heat dissipation, especially during the summers on the fringes of the tropics. This literature proposes new ideas and innovations that consider the ultimate Beescape for Meliponaries, with bee comfort at the top of the list of considerations.

Heat waves in the tropics may leave beekeepers anxious about the stability of the colony's health and population growth. Chances are the existing colonies may see the population dwindling rather than growing at any rate. We also look at earth tremors, if not earthquakes, in the "ring of fire" zone in The Malay Archipelago and Wallacea regions. Flooding threats in Southeast Asia and Typhoons in the Oceanic Philippines are troubling. Recent reports on landslides leave affected beekeepers very distressed.

Addressing these disastrous events and the potential threats is one of the aims of this book. We have scoured the region and examined how different cultures attempt to mitigate their regions' drastic and extreme weather. Out of these examples and data collection, we provide a gallery of designs on relevant bee housing that may impact such mitigation that may apply to the relevant type of disaster frequently faced in each region.

The reader is afforded a choice of constructive possibilities depending on the availability of materials, site topography, and geography of their Meliponary location from the Indian Subcontinent and the Himalayan States through ASEAN countries.

Acknowledgements

The Author expresses gratitude to Ar. Dr. Mastor Surat. Dip. Arch., BArch. (Hons) (UTM), M.Litt. (UKM), PhD (UKM) P. Arch LAM, APAM, AIPDM, for his advice on the structure of this book.

We are much obliged to our mentor, Dr. David Ward Roubik of the Smithsonian Tropical Research Institute, Panama, for all their extremely useful and indispensable advice and guidance towards preparing major portions of this book.

The author is very grateful to Dr Claus Rasmussen, the Danish entomologist. Department of Agroecology, Aarhus University, Denmark, for his advice and for sharing the Type Locality of the many Meliponine species in this Indo-Malayan/Australasian region.

We are thankful to Dr. Mohd Affendi Mohd Shafri, Associate Professor, Dept. of Biomedical Science, & Project Leader of Malay Medical Manuscript Flagship Project, International Islamic University Malaysia, for contributing to the Preamble.

We thank Mr. Mohd Razif Mamat of the Malaysia Genome & Vaccine Institute for his resources and complete support, without whom this book would not have materialised.

Meliponine Type depository abbreviations used in the listed Records.

AMNH - American Museum of Natural History
AMS - Australia Museum Sydney
ANIC - Australian National Insect Collection
BMNH - British Museum of Natural History
BPBM - Bernice P. Bishop Museum
Calicut - Calicut University, Zoology Department. Kerala/ India
DEI - Deutsches Entomologisches Institut im ZALF. Müncheberg, Germany
EKYU - Entomological laboratory, Kyushu University, Fukuoka, Japan
HNHM - Hungarian Natural History Museum,
IEBR - Hanoi, Institute of Ecology and Biological Resources, Vietnam,
MNHN - Muséum National d'Histoire Naturelle, Paris, France
MSNG - Museo Civico di Storia Naturale "Giacomo Doria"(Collezione Gribodo), Genova, Italy
MVMA Australia, Victoria, Abbotsford, Museum of Victoria (Ken Walker)
MZB - Museum Zoologicum Bogoriense, Bogor, Indonesia
OUMNH Oxford, University Museum of Natural History, United Kingdom,
QM - Queensland Museum, South Brisbane, Australia,
RMNH - Nationaal Natuurhistorische Museum ("Naturalis"), Leiden, Netherlands
SEHU - Systematic entomology Hokkaido University Museum, Sapporo, Japan
SEMC Snow Entomological Museum University of Kansas, Lawrence, Kansas, USA

Book cover design description: The Batak Karo house is typically from Northern Sumatra and is normally rectangular; however, it is also found in West Kalimantan. An isolated case, but interestingly, the Batak Karo traditional roof is built on a Dayak tribal tall oval house on very high stilts. It keeps occupants safe from wild animals and floods. Redesigned with half the stilts' height and the roof with a 75% reduced height. Visually compressing vertically, the whole structure makes it more realistic and appropriate as a bee hive shelter. Even so, reduced overall height can still provide three levels of usage. This also addresses the issue of resource partitioning for the bees. At the ground level, the resources are limited to shrubs and groundcovers. However, the possibility of hive placement at higher levels provides easy forage of the tree canopies.

Figure 1 Vernacular Fusion – Batak Karo roof on a West Kalimantan tall house. The original site stands in Darit, Menyuke, Landak Regency, West Kalimantan, Indonesia

This image was chosen as the book cover because it covers most of the features proposed in this book. It is a fusion of two major cultures in the Greater Sunda Islands of Sumatra and Borneo on the Sunda Shelf. Geographically, they are the centre of the Indo-Malaya ecozone.

The alternative bee box hive with the roof design looking conical seems applicable for a hollowed-out log nest, and the monitor roof is detachable for observation of the nest internals. This design can still be applied with a monitor roof for observation hives.

Contents

Sustainable Meliponiculture with Vernacular Architecture 1
Volume 1 - The Malay Archipelago 1
  Synopsis: 1
  Acknowledgements 2
  Book cover design description 3
A Preamble 8
Preface 9
PART 1 11
~ MALAY ARCHITECTURE ~ 11
Prelude to Part 1 12
  Miniature Model Making 12
  Introduction to Part 1 14
Chapter 1 15
  Conservation in Meliponiculture 15
  Bee housing in Meliponiculture 16
    Vernacular Architecture for Bee Housing in Meliponiculture? 16
Chapter 2 17
The Malay World 17
  Historical origin 17
    Early conception 17
  Malay Vernacular Architecture 19
  Main Features 19
  Classic attributes of Malay tradition 22
    Flexibility 23
    Function 23
  Aesthetics 24
  Types of Malay States Traditional House 25
    1. Perak House 25
    2. Selangor House 27
    3. Melaka House 27
    4. Penang Island House 29
    5. Kedah House 31
    6. Negri Sembilan House 31
    7. Pahang House 32
    10. Perlis House 34

Rumah Panjang Kadazan Sabah or Sabah Kadazan Longhouse 39
   Adaptation of Vernacular structures for Meliponary designs.. 43
      The Melanau Traditional House 44
   Malay Culture, Architecture and Literature 44
   Translated excerpts from https://ms.wikipedia.org/wiki/Kebudayaan_Malaysia................................... 44
   Bangka-Belitung Islands 46
      Rumah Brunei.. 46
   Vernacular architecture and Insect Tourism 47
   The traditional house turned Bee Gallery by AHJ. 50
   Box hive roofs 52
   Min House Camp, Kubang Kerian, Kelantan by AHJ... 54
   Big Bee Honey Gallery, Maran, Terengganu 56
   by Mohd Razif Mamat... 56
      Vernacular architecture in the Malay Peninsula. 57

Chapter 3 58
   Malay Nusantara traditional architecture 58
      Model making among beekeepers 62
   Malay Heritage Centre (MHC) in Singapore 63

Chapter 4 65
   Sustainable Meliponiculture in Brunei 65
      Meliponiculture in Tutong, Brunei by Mitasby Amit... 65
   Urban Meliponiculture in Bandar Seri Begawan – 67
      Meliponicultursts (Stingless Beekeepers) in Brunei Darussalam.. 68
   Wild Bees of Brunei Darussalam David Roubik 1996v 70
   Reference:.. 70

PART 2 71
~ INDONESIAN ARCHITECTURE ~.. 71
   Introduction to Part 2.. 72

Chapter 5 73
   The Bataks of Sumatra 73
   Batak Karo of North Sumatra 73
   Batak Gayo 75
   Gayo Lues tribe in Gayo Lues district 75
   Gayo Kalul in Aceh Tamiang 76
   Batak Bebesen in Aceh Tengah 76
   Batak Gayo Serbejadi in Aceh Timur 77

Batak Gayo Deret in Aceh Tengah 78

Batak Gayo Lut in Bener Meriah 78

Bataks in Aceh Province 79

Distribution of Meliponiculture with Vernacular architecture 82

Chapter 6 83

Aceh and the North Sumatra Architecture 83

The Aceh Escapade 85

Minangkabau of West Sumatra 86

Chapter 7 89

Vernacular architecture in Central and South Sumatra 89

Malay Lontiak House of Kampar Majo Tribe 89

Sesat Balai Agung 92

Rumah *Nuwou Balak* (Rumah *Kepala Suku*) 92

Rumah Adat Lampung *Nuwou Lunik* 93

The Meaning of the Parts of the Lampung Traditional House 93

Enggano Island, off Bengkulu Province, SW Sumatra 94

Chapter 8 95

Javanese Architecture in Meliponiculture 95

Sundanese Architecture of West Java 100

Rumah Adat or traditional house 102

Kasepuhan House of West Java 103

Central Java 104

Bawean Island, off Gresik Regency, East Java 105

Chapter 9 106

Traditional Architecture of Lesser Sunda and NTB- 106

Balinese architecture 107

Vernacular Architecture of Lombok Island 109

Chapter 10 112

Meliponiculture Tourism & Ancestral Veneration in Ethnic Cultures of Sulawesi 112

Gorontalo Province 113

Vernacular architecture of Sulawesi 117

Chapter 11 120

Vernacular architecture in Kalimantan 120

Kalimantan Vernacular Architecture Chart 125

Chapter 12 127

Meliponiculture in Indonesian Wallacea (NTT region) 127

    Vernacular architecture of Sumbawa 127

  Vernacular architecture of Maluku Islands 129

    8.   Buru Island - 132

    9.   Halmahera Island 133

Chapter 13 135

  Vernacular architecture of East Nusa Tenggara 135

  Flores Island 135

  Komodo National Park 140

  The Sama Bajao Factor 141

Chapter 14 142

  Vernacular Architecture of Sumba and NTT group of Islands 142

  Letmafo Society, North Central Timor District 144

  Group of Solor Islands 144

  History of the Solor Islands 145

  History of the kingdoms of Adonara island 145

  Lembata Island 148

  Savu & Raijua 148

  Pantar Island 149

Addendum Gallery 152

  Ideas Gallery for Vernacular Bee Housing, Racks, Shacks & Sheds designs. 152

List of Figures 156

Index 164

  Bibliography 168

A Preamble

Bees and Southeast Asian cultures cannot be separated. There is so much in the literature that points to this. The manuscript *Peringatan Raja-raja Kedah* talks about how beeswax or cerumen (stingless bees mix wax with resinous materials to make the latter) was a key trade of the Sultanate of Kedah. The harvest of beeswax from 'Tualang' bees (giant honeybees) was for the Sultan, who reserved the exclusive right to sell it. Beeswax collection was so important to Kedah that whenever a particular Tualang tree bore beehives, the royal tax man would come to estimate the duties and a special post called Penghulu Tualang would be appointed for that particular tree to oversee its collection. Honey from these local bees is also frequent in various Malay medical manuscripts. I have studied these manuscripts in my search for new drug therapeutics. I found, for example, in Bustan al-Salatin, written under the order of the Sultan of Aceh, that honey is deemed useful for clearing phlegm in the lungs and stomach, providing good vision and strengthening the body in general. The royal *tabib* of the Sultans of Pontianak recommended using honey to treat dysmenorrhoea and post-natal contracting pain. Other manuscripts of traditional Malay medicine tell us to use honey to treat problems like headaches, sore throat, numbness of feet, rheumatism, and bone ulcers. Honey in Malay medicine is seldom used on its own and is usually diluted as a watery preparation consumed as a drink or added to a concoction of other medicinal ingredients.

Bees and honey thus inspired my research on anti-pain, anti-fever and anti-inflammation. Considering the numerous more stingless bee species, I came to appreciate honey bees' different types and qualities and the importance of apiculture and Meliponiculture. Different factors – temperature, availability of food, and housing - provide for the different grades of honey even if the same type of bees is used. Hence, I could relate to the value of this book's idea.

I must say that I also love how science, architecture and culture intertwined nicely in this book. Incorporating the local architecture of the regions into bee houses seemed like a very good plan at many different levels. So many creative designs are showcased in this book to represent the various sub-cultures of the Southeast Asian region.

We hope there will be an improvement in the bees' quality of life as much as there will be a boost in awareness of each local indigenous culture celebrating bees and honey throughout the ages. Paraphrasing a saying purportedly by Marcus Aurelius: That which is good for the beehive should be good for the bees.

Mohd Affendi Mohd Shafri,
Associate Professor, Dept. of Biomedical Science,
& Project Leader of Malay Medical Manuscript Flagship Project,
International Islamic University Malaysia.

Preface

Recollecting chats with Dr. Claus Rasmussen about Meliponary structure designs, hoping to get some insights because he had a Meliponary in Peru many years ago. He said, "I have had several different designs. In general, hives are only kept 1-2-3 or a few, stored in the house. Therefore, there is no particular meliponary design." We agree that no specific design can lend to a "standard" Meliponary design. However, we can add local flare or indigenous ambience to any structure. Utilizing locally available materials and natural fibres from the surroundings would give significance to adaptation to the regional habitat. This will add curiosity and attraction to the insect tourism sector.

This book is written for Beekeepers. The idea is to show what has been done and practised in different regions. Each region has peculiar weather extremes and different cultural habits of the bees and their keepers. People can then choose or try out different designs for the comfort of their bees. It also intends to include innovations and new ideas in line with each culture's identity and heritage. These innovations, in turn, will help those also thinking of tourism. The other thing is facilitating bee sanctuary hopefuls to start their sanctuary structure development and management. When considering the bee's comfort, we have realised that the bee housing issue may be a bit out of the normal bee scientist wavelength.

On the question: why are honeycomb cells hexagonal? Prof. Alan Lightman of MIT says, "There is often an energy principle operating in nature. Outcomes that require less energy tend to be favoured. The answer includes all principles, mathematics, economy, and the least energy cost. Only three shapes with equal sides can be attached to maximise quantity over space. Squares, equilateral triangles, and hexagons. They could try other irregular shapes, but the problem is to have each cell custom-fit the next one. [Ed.: However, stingless bees make round individual brood cells, beginning in the middle of each horizontal comb under construction, and the first cell is round, not hexagonal. Cells become hexagonal as more cells are built around them, pushing in their sides (Roubik, 2012, 2018)]. Otherwise, each bee must wait in line to build the honeycomb. Each bee must wait before the previous one completes its cell to fit the previous one without leaving gaps because gaps will waste space, introduce bacteria, [Ed.: be more vulnerable to any invasion or predation], etc. So, to build things sequentially, all the bees are working simultaneously. Of the three shapes, the hexagon has the least total perimeter, meaning that less material is required for construction. Every ounce of wax produced requires some eight ounces of honey. [Ed.: Bees make their wax in glands and excrete it as tiny scales or plates over parts of their abdomen, from which it is removed and moulded into nest material.] As the bees like to economise and save energy, they use the hex to cover a given area."

Exploring the hexagon from that kind of inspiration, we look at cabanas, beach huts, gazebos and even Chinese pavilions in the hexagon form. Standing in the middle of a six-sided structure, one can have a 360º surround view. Even the six pillars give very few blind spots. Some bee farms have adopted this advantage to place their box hives on hexagonal racks, allowing them to fly in and out in six directions.

This predicament inflates when reminiscing about a World Science Festival program. The program moderator was John Hockenberry, and the participants were Robbert Dijkgraaf, David Gross, Alan Lightman, and Maria Spiropulu on June 4, 2016.

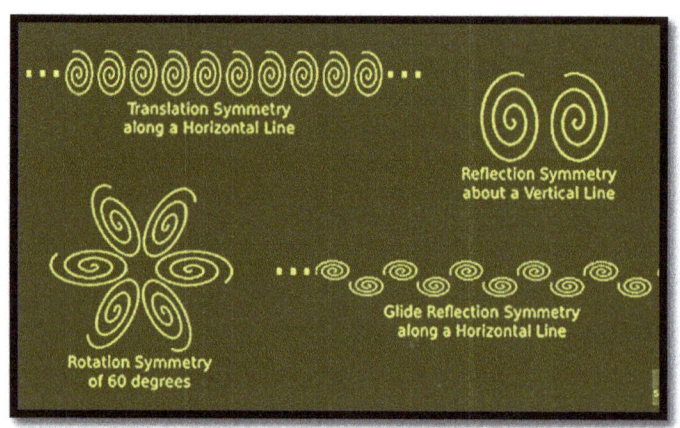

Figure 2 Beyond Reality – Power of Symmetry. Source: https://www.youtube.com/watch?v=X6HobTJ2jnk

Reflecting on Robbert Dijkgraaf's illustration of Translation symmetry, one is reminded of the Native longhouses and how beekeepers have subconsciously adopted this translation symmetry in their bee racks design.

From a bee's hexagonal honeycomb to the elliptical paths of planets, symmetry has long been recognised as a vital quality of nature. Einstein saw symmetry hidden in the fabric of space and time. And today's theorists are pursuing an even more exotic symmetry that, mathematically speaking, could be nature's final fundamental symmetry: supersymmetry.

In the early 1960s, my family lived in Penang Island in the Govt. quarters on the same road as a Jewish cemetery (which turned out to be the oldest Jewish cemetery in South East Asia).

Figure 3 Rolled-up Bamboo blinds with a SB nest within.

There, as a child, I discovered the bees (not knowing then they were stingless bees). I found them in Jewish and Chinese tombstones and graveyard perimeter walls. In school, the bees were on branches, trees in the school compound, and structures in Botanical gardens and beachfront huts, even in the rolled-up blinds and in cracks of old door jambs. This phenomenon intrigued me to this day, and I decided to write about bees and vernacular architecture.

Abu Hassan Jalil, Malaysia 2022

PART 1
~ MALAY ARCHITECTURE ~

Prelude to Part 1

Miniature Model Making

En. Baser Ali of Kg. Morten, Melaka, is a craftsman whose profession is a master carpenter by day and a miniature house model maker by night. Below are some of his works displayed with explicit permission with the hope of aspiring beekeepers building the same to house stingless bees to keep the industry flourishing.

Figure 4 Assorted Miniature models and bottom right: En. Baser Ali at work

Miniature Model Making

Figure 5 More models and the bottom is the Melaka Palace replica

Introduction to Part 1

Part 1 starts with the conservation of stingless bees or Meliponines. These insects are national treasures of each country in the Indo-Malayan ecozone. They provide health and wealth through their products and derivative by-products. Meliponiculture is currently a valuable resource with the marked decline of Apiculture globally. Unlike Apiculture, there is no serious disease experienced in Meliponiculture. Only a matter of controlling and avoiding pests and predators. This can be achieved with proper conservation in Meliponiculture.

The next chapter looks at vernacular architecture for bee housing in Meliponiculture. The reason for looking at vernacular, indigenous or ethnic architecture is because of this indigenous knowledge of air movement, ventilation and shelter that provides thermal comfort in the tropics.

Vernacular architecture is a major theme in cultural and heritage village parks. Many miniature model gardens are tourist destinations in many ASEAN countries. The Philippines has its heritage parks, and Indonesia has Taman Mini Indonesia. Also, we have Taman Mini Malaysia and even Taman Mini ASEAN in Malaysia. Both Sabah and Sarawak have their own culture and heritage parks. Having bees in these miniature or full-scale model houses can only enhance insect tourism as an added feature in these cultural parks. I dedicate full chapters to regional indigenous architecture, including Traditional homes that have been converted into Bee Galleries. These galleries house bee colonies and bee products like honey and pot-pollen plus derivative products.

As an added feature of Meliponiculture tourism, this literature covers a passage on ancestral veneration in ethnic cultures across the Indo-Malaya region. This ancestor veneration culture includes spiritual beliefs and even shrine architecture design and worship, e.g., *puja* in Balinese culture.

This part also looked at some box hive designs in Australia for comparative ideas. Frequent hive design exhibitions move innovation forward among Australian beekeepers. Regarding Meliponines, Australia has about eight or nine native species. Hive box designs need not be too elaborate as they are similar in size and habits. They have some classic beehive designs and also elaborate cylindrical hive containers. The designs reflect an adaptation to the seasonal climate of the southern tropical and subtropical peculiarities. Unlike the equatorial tropical weather, Australian designs are made to adapt to the winters.

Chapter 1

Conservation in Meliponiculture

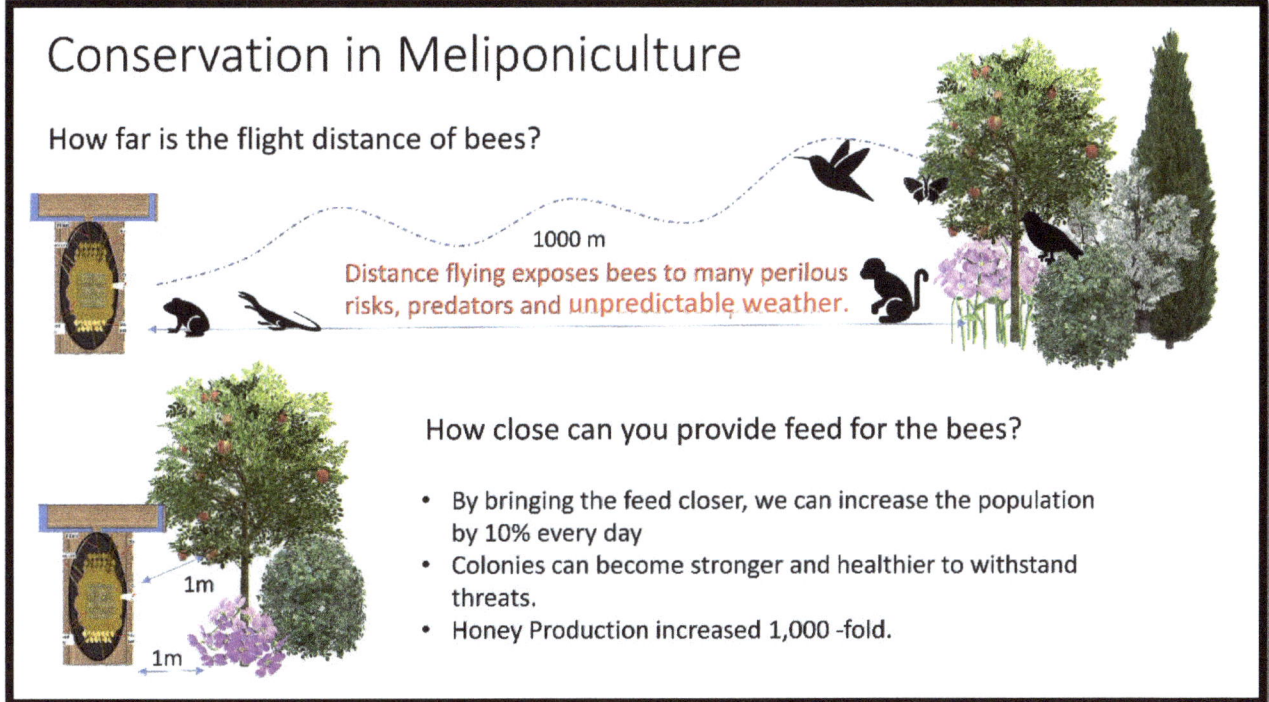

Figure 6 Conservation in Meliponiculture

Bee scientists, entomologists and researchers frequently scrutinise the bees' flight range and forage distances. This should not be crucial information for a beekeeper. The keeper's more important concern should be how close they can provide foraging sources. The further the distance the bee travels, the greater its dangers. Besides predators and competitors[1], they face extreme weather changes like a sudden downpour of strong gales that can divert their return path or the scent trail of their forage destination. These conditions also reduce the productivity of the whole colony.

The other factor to consider is the daily death rate and losses in the colony population. Assuming out of every one hundred bees fly out to forage, eighty may return consistently, but ten would reach their last flight at the end of their life, five may lose their way back, and predators may defeat the rest. We cannot save the ten at the end of life, but the five that lose their way back and those that were preyed upon may be saved by bringing their foraging needs as close as possible. (See Figure 6)

---

[1] Editor's note: In theory, there are scramble and contest competitors, and bees do either or both. Whoever gets there first, or whoever defends the resource, is the "winner"- in the short term. In the long term, since the resource is a plant that requires, most often, a bee to pollinate it, losing to a competitor can later be advantageous when all the seeds it has produced as a pollinator produce plants that have flowers and feed the "enemies"/ Don't be fooled.

Bee housing in Meliponiculture
by Abu Hassan Jalil

Vernacular Architecture for Bee Housing in Meliponiculture?

Living in the tropics has its challenges with humidity and heat. When I did my technical drawing course, there was a subject on Architectural design. I was always intrigued by how the older generation got by without air conditioning in wooden houses. My grandparent's house was always cool and did not need electrical devices to provide extra comfort. It turns out that it was all in the architecture.

With this knowledge[2], I frequently encourage fellow beekeepers to construct traditional roofs for their box hives for better heat dissipation and to maintain their unique identity. This identity I did by publishing my first book on Beescape in 2014 with some traditional architecture on the back cover.

Figure 7 Examples of Bee housing with vernacular roof designs. Image from the back cover of Beescape Book 2014

These models were built ten years ago with friends in North Peninsula Malaysia. The Malays had always been great woodcarvers and excellent skilled carpenters in their own right. Early examples were the Malaccan Double pitch, The Javanese Joglo, and The Minangkabau curved ridge. It took some time and much nudging to others in the Nusantara region through social media interest group platforms to catch on to this trend.

After ten years, it is exhilarating to see that the trend has picked up and evolved from the minimalist Gable roof to many innovations after ten years. Even miniature model hobbyists 'jumped on the bandwagon'.

---

[2] In our hot and humid climate, our traditional Malay house is the best house design comparing to other traditional houses. Built and design by the local craftsmen, every inch has its soul. Every construction has its function not only for aesthetic value. http://awesomelypatiencelikewonderwomen.blogspot.com/2014/01/sustainability-traditional-malay-house.html

Chapter 2

The Malay World

■ Countries most often considered to be part of the Malay world (Brunei, Indonesia, Malaysia, and Singapore)

■ Countries that have historically been influenced by Malay culture (the Philippines, Sri Lanka, Thailand, and Timor-Leste)

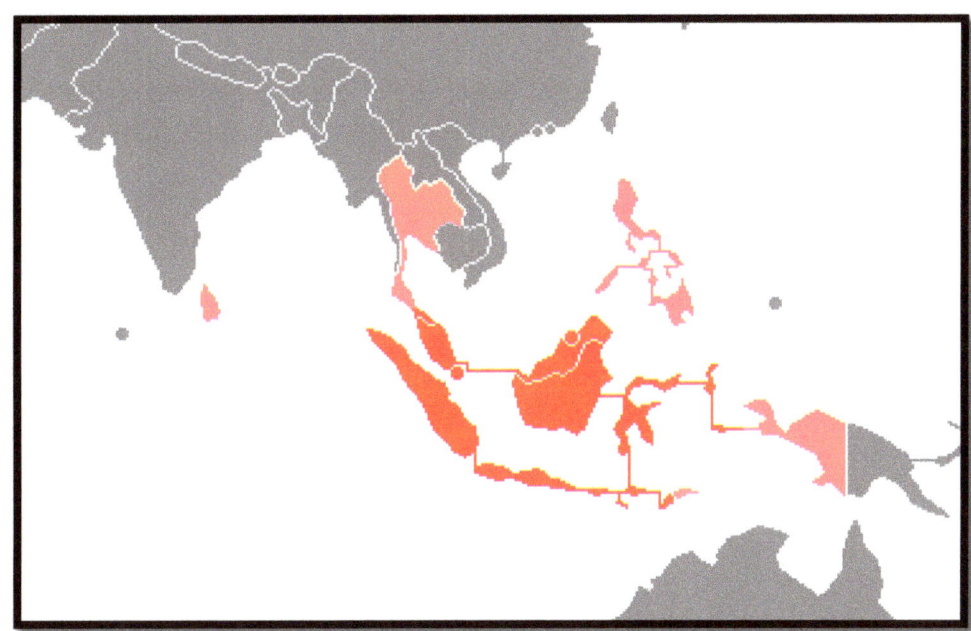

Figure 8 The Malay World in the Indo-Malaya Ecozone

Historical origin

Early conception

Main articles: Sumatra, *Melayu* Kingdom, and Srivijaya

"... starting point by the Island of Pulo Catay in the region of Pattane (Pattani), situated in the east coast in 8 degrees of latitude, the pass round to the other or western coast of Ujontana (Malay peninsula), to Taranda and Ujon Calan situated in the same latitude in the district of Queda (Kedah): this stretch of the territory lies within the region of "Malayos", and the same language prevail throughout ..."

– Manuel Godinho de Erédia, 1613.

Territorial identification of Malay is of ancient origin. In the 7th century, the term gradually became an ethnonym throughout the consolidation of Srivijaya as a regional power. Various foreign and local records show that Melayu (Malay) and its similar-sounding variants appear to apply as an old toponym to the ancient Sumatra region in general. Tomé Pires, an apothecary who stayed in Melaka from 1512 to 1515 after the Portuguese conquest, explained how the former Malacca classified merchants calling its port into four groups, of which the Malays or Melayu did not appear in the list, suggesting they were not then regarded as a category outside the Melaka itself.

Another term, Malayos or the 'Sea of Malayu', was espoused by the Portuguese historian Manuel Godinho de Erédia to describe areas under Malaccan dominance. The area covers the Andaman Sea in the north, the entire Straits of Malacca in the centre, a part of Sunda Strait in the south and the western South China Sea in the east. It was generally described as a Muslim centre of international trade, with the Malay language as its lingua franca. Erédia's description indicates that Malayos were a geo-religio-sociocultural concept of geographical unity characterised by common religious beliefs and cultural features.

An identical term, Tanah Melayu (literally 'Malay land'), is found in various Malay texts, of which the oldest date back to the early 17th century. It is frequently mentioned in the Hikayat Hang Tuah, a well-known classical work that began as oral tales associated with the legendary heroes of the Melaka Sultanate. In the text, Tanah Melayu is consistently employed to refer to the area under Melakan dominance. In the early 16th century, Tomé Pires coins an almost similar term, Terra de Tana Malaio, for the Southeastern part of Sumatra, where the deposed Sultan of Melaka, Mahmud Shah, established his exiled government.

"... the country which Europeans denominate the Malay Peninsula, and which, by the natives themselves, is called 'the land of the Malays' ('Tanah Melayu'), has, from its appearing to be wholly occupied by that people, been generally considered as their original country ..."

– John Crawfurd, 1820

The application of Tanah Melayu to the Malay Peninsula entered the European authorship when Marsden and Crawfurd noted it in their historical works published in 1811 and 1820, respectively. Another important term, the Malaya, an English term for the Peninsula, was already used in English writings from the early 18th century.

Due to the lack of available research, it is difficult to trace the development of the concept of the Malay world[3] as a term that later refers to the archipelago. However, thus, classical territorial identifications are believed to have formed an important antecedent for the future conceptualisation of the Malay world. The term "*Alam Melayu*"(Malay World) did not exist before the 20th century. Classical Malay works of literature between the 14th century to the 20th century never mentioned "Alam Melayu" or any similar term. Instead, the term emerged along with the emergence of the Malay identity and nationality movement after 1930, mentioned in Malay periodicals such as *Majalah Guru* magazine, *Saudara* newspaper, *Majlis* newspaper, and *Puisi-Puisi Kebangsaan* newspaper.

---

[3] https://en.wikipedia.org/wiki/Malay_world

Malay Vernacular Architecture

Main Features

All nations and peoples have their own traditional house, the knowledge of which is passed down from one generation to the next. People who sit in a cool area have their home designs. People in hot areas also have their own designs.

The Malays are also great at building this traditional house. Even though times have changed and our homes are getting more and more storied and concrete, the Malay Traditional House still exists and may be a status symbol. Knowing how to take care of this wooden house is difficult.

1) Always cool

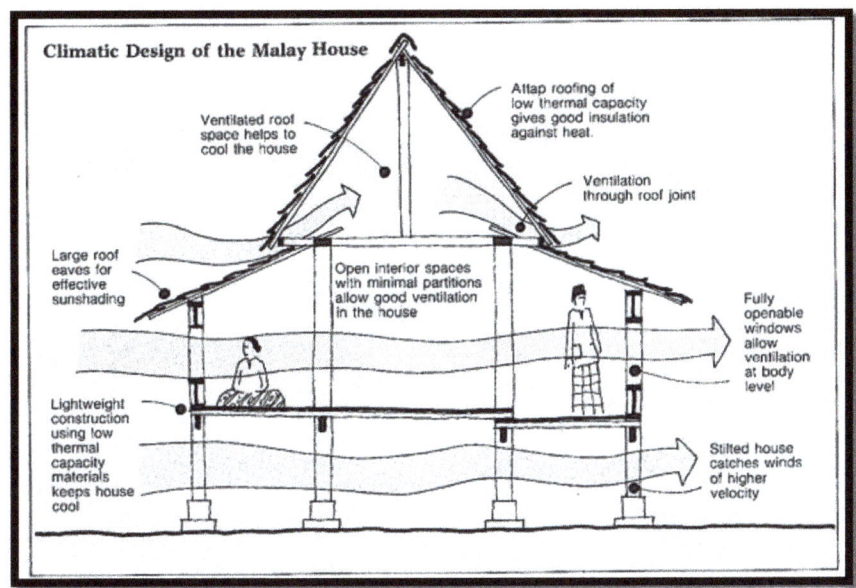

Figure 9 Passive ventilation and airflow in a Malay traditional house

You will feel cool and peaceful as you enter any traditional Malay house. What more at night? Not surprisingly, this is because the village house has a lot of air movement. This passive ventilation is applied in designs on the floorboard, above the window, or roof. The vents aren't just for fun; they have airflow functions, allowing air circulation.

2) Tall houses or stilt-raised platforms (*Rumah Panggung*) are safer

Village houses are always built on stilts. Not only on whim or fancy but it was also built on safety factors, such as avoiding being raided by enemies and being entered by thieves, avoiding wild animals and natural disasters such as floods.

Logically, modern construction looks intact in terms of durability, but it cannot adapt to the climate and weather of a place. This is an advantage that can be seen in traditional Malay houses. The high-rise floors that are part of the identity of a traditional Malay house not only offer safety protection for its occupants from the threat of enemies and wild animals.

When the wind enters from under the house, it can enter through a floor installed with sparse floor timbers. This condition is ideal for helping the process of conversion and air circulation. It can also maintain the internal thermal comfort of a house.

3) Entrance Platform or *Anjung*

The *Anjung,* or veranda, is the main resting place of the village house. It symbolises how the Malays attach immense importance to the customs and order of society.

It is also to avoid slander; male guests are only invited to the pavilion. There is a separation of space so that social life is peaceful. The Anjung is the platform beyond the portico[4] or porch before the main entrance door.

4) Wood carving or *Ukiran*

One may observe that carving is widely available on railings, stairs, walls of verandas and roofs of village houses, proving that the Malay community is reliable in carving and carpentry.

Not just aesthetics, this carving promotes ventilation and lighting for the home. The room is closed but still associated with the outside environment. Form and function in a barrel.

2) Pediment[5] or *tebar Layar* of the Kelantan Twelve Pillar House *Rumah Tiang Dua Belas*

In Kelantan, the traditional house is known as the Twelve Pillar House, resulting from the country's unique and long history and background.

Based on the name, we know that this house has 12 pillars! It has six portico poles and six long pillars in the main room.

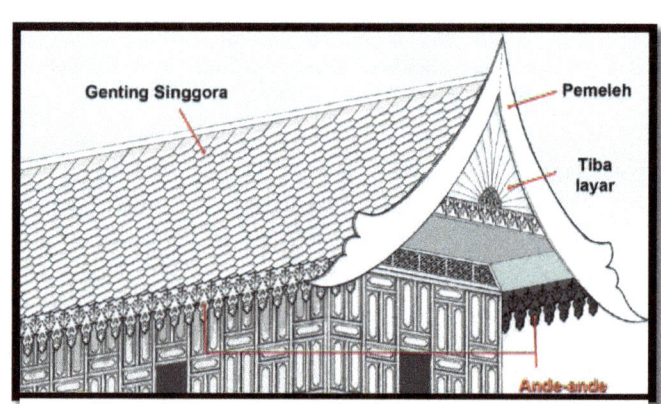

Figure 10 Typical parts of a Malay vernacular roof

The main space is divided into three parts, namely the *Rumah Ibu* (Mother), *Rumah Tengah* (Middle) and *Rumah Dapur* (Kitchen). *Rumah Ibu* has a spacious living room to serve guests. The *Rumah Tengah* (Middle) also includes a bedroom, living room, and a *Rumah Dapur* or Kitchen for cooking.

---

[4] Portico - A portico is a porch leading to the entrance of a building, or extended as a colonnade, with a roof structure over a walkway, supported by columns or enclosed by walls

[5] Pediment - the triangular upper part of the front of a classical building, typically surmounting a portico.

Interestingly, this Twelve Pillar House is the design of the *pemeleh* or rake board and the mast. What is that? These pickets are two pieces of board placed at the top of the end of the screen, like in the picture below.

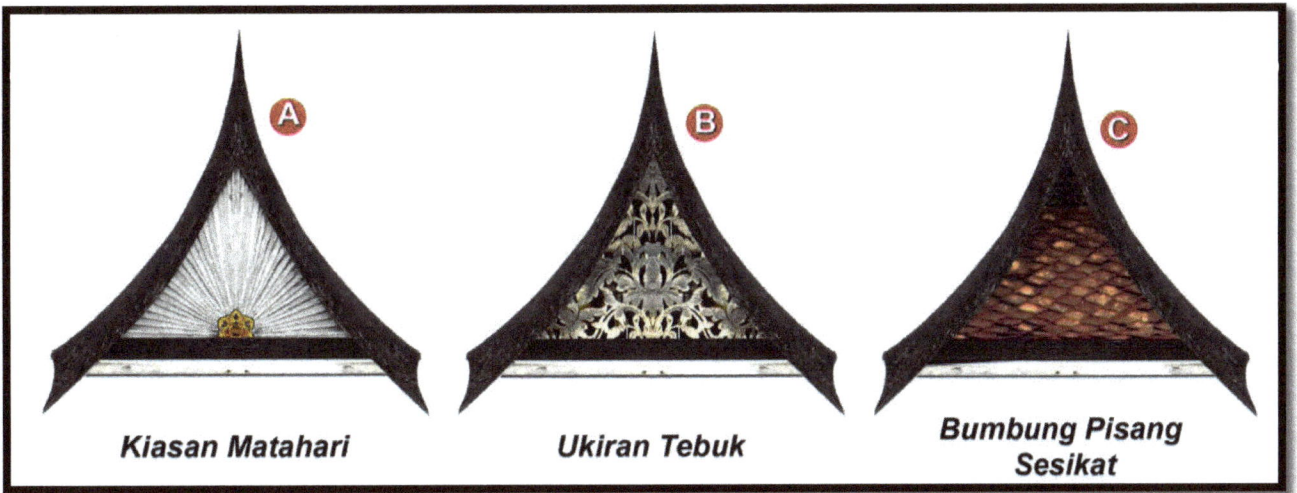

Figure 11 Various rake boards for a Kelantan House

There are many meanings for this *pemeleh*; some say it is the symbol of a stork whose carvings are widely found in local people's canoes, and some say it is related to the Islamic faith. In short, it is a function for each design. An implied purpose of life is applied in the construction of this house.

The carving on the screen is also incredibly beautiful. The cut-through motifs allow natural light in and provide ventilation. In addition, the fascia boards called ande-ande are also ornately carved. This is one of the inspirations for traditional houses that we can apply in modern homes.

4) Privacy in *Rumah Limas* Johor

The architecture of the Johor *Rumah Limas* is heavily influenced by foreign cultures, including Dutch and British colonialism in the Malay Land, as well as the influence of Islam and Indonesian architecture (See Part 2 of this volume). Limas or "*bumbung lima*" (five roofs) have the same basic form. It is built with a simple height. It also has stairs like other traditional Malay houses.

The Limas Johor house is divided into the front, middle and rear. However, the Johor Limas House can be recognized by its long perabung (ridge) roof, which is connected to four pyramidal roofs. The seclusion of this room is to maintain the host's privacy but not to marginalize the guests.

For example, the front is devoted to serving guests, *marhaban*[6] and "showing off" the prosperity of the host's life. These Malays have always been good at protecting privacy!

---

[6] Marhaba (Welcome) It comes from the word *"rahhaba"* which means "to welcome".

5) Clever design in Pahang *Rumah Serambi* or Porch House

The Serambi Pahang house usually has a long roof. Its uniqueness can be seen through the construction technique with wood and without using any nails on the walls and floor.

What's interesting about this Pahang Porch House is the extra form of *rasuk* as a binder to the house frame. The goal is to add space later for future expansion. Suddenly, the family grew, using the existing *rasuk* to add space to the house.

In addition to having three main rooms, namely the front porch, the 'mother' or *Rumah Ibu* and the kitchen porch, the house's uniqueness also lies in the "geta" or raised floor for certain purposes, including as a place to rest and sleep.

The Serambi Pahang house also has a barrier known as a *bendul* or threshold about 0.4 meters high, which is the entrance from the 'mother' house to the porch.

(Ref: https://www.propertyguru.com.my/bm/panduan-hartanah/inspirasi-rumah-melayu-tradisional-40051 Tramizi Hussin accessed 8 May 2021)

Classic attributes of Malay tradition.

In the past, Malay houses were usually not cubicles because the Malay community at that time emphasized meeting or gathering in groups for public majlis and religious ceremonies, 'motherhouse', cubicle, inner porch, kitchen platform or *lambur*[7] and clothesline as well as the basement of the house.

Figure 12 Rumah Lipat Kajang, a type of traditional Riau house with tiled stairs in the Taman Mini Indonesia theme park. Source: https://www.wikiwand.com/en/Malay_house

From the design point of view, the Malay house can be said to be very sophisticated and contains the *bestari*[8] characteristics, again in contact with a comfortable lifestyle according to customs and nature. These smart features include natural ventilation, shading, safety, security, and 'privacy'. For example, the most obvious characteristic of a traditional Malay house is that it is built on stilts. Various advantages arise from this 'float', including avoiding the easy entry of wild animals and being lifted from the flood level. The

---

[7] A kitchen platform for preparing ingredients. Spices and condiments for cooking
[8] Bestari - (poetic) smart, well educated.

house is also cooler because of airing on all sides (from the top, edges and bottom). In addition, the space under the house is also called 'pods', which is a suitable place to store all kinds of utensils.

The various carved cut-throughs on the walls are an aesthetic accent and permit air and light to enter the house. The carvings on Malay houses have elements of geometric or plant patterns (there are no living fauna patterns due to religious prohibition), and many are very richly carved. There are various variations of the windows of Malay houses according to the tastes and skills of the past craftsmen. The windows dominate the view from all angles of the house and provide adequate airflow to the house.

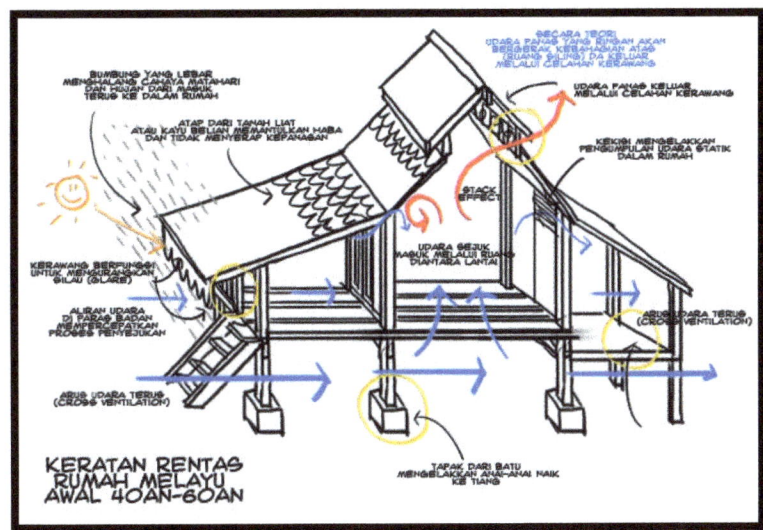

Figure 13 Cross-section of a Malay House in the 40s to 60s.

We shall attempt to translate these features into the Beehouse designs.

Flexibility

The home garden, or the house's front yard, is the first room in the compilation of a Malay landscape. The front room of the usual room has a spacious courtyard area. The situation is sandy and clean of garbage and grass. The house's front yard is roomy because it functions as an area for ceremonies and social activities. The activities usually carried out are wedding ceremonies and traditional games such as tops, long poles, martial arts, etc.

The flowers planted in the urn are arranged around the front yard. Planting flowers in pots principally is for plant transfer. This situation is flexible regarding its location to allow the execution of activities in the yard.

Function

Plants cannot be separated from the Malay community. It closely relates to life regarding customs and the community's way of life. (*Tamadun Melayu*[9]; Ismail Hussien et. Al., 1995)

---

[9] Tamadun Melayu Jilid I - 4/ Published: Kuala Lumpur: Dewan Bahasa Dan Pustaka, 1993 - 1995.

The role of plants can be divided into two, namely:

1. Immediate use

a) main food source

b) household necessities and equipment

c) traditional medicine

d) cosmetic plants

e) weapons and poison materials

2. Indirect use

a) enriching literary treasures

- rhymes - proverbs - songs - terms - traditional ceremonies - have mystical powers

Aesthetics

It is part of Malay art. In this aspect, aesthetics refers to the value and feeling of something beautiful, isolated from its function. The emphasis is more on the artistic aspect. Malay art can be expressed in all art forms, such as carving, metal art, rhymes or poems and other artistic activities.

Figure 14 Typical Terengganu House

This Malay art form is evidenced in the classic Malay literature "*Hikayat, Syair Burung Pungguk*". Himayat means a story, tale, narrative, or narration. Syair is a poetic form of a saga recitation without the accompaniment of musical instruments. This form includes pantun and *Gurindam*[10] and folk storytelling in melodic verbal expression.

Using images in the pantun proves the aesthetic image in Malay society. The rhymes and poems also use nature images as shading materials, such as trees, flowers, animals, seas, and rivers. This shows that there is a connection between nature and humans. Following the reality in Malay cosmology, which is related to God, nature, and humans.

Subjects: Malay (Asian people) -- > Congresses Culture -- > Congresses; Malay Archipelago -- > Civilization -- > Congresses

[10] Gurindam (Jawi: كور ⌧ ندام) is a type of irregular verse forms of traditional Malay poetry.

Types of Malay States Traditional House

1. Perak House

Figure 15 Inspired by the Rumah Kutai or Rumah Warisan of Perak, Malaysia.

Figure 16 In Perak, the long-roof house is known as the Kutai House and Rumah Potong Pattani.

This house is of the long roof type. In essence, this design is that it is more ethnic and concise. This design is also known as the Kutai house. These houses are still in the state of Perak and are over one hundred years old. These include Kampung Sayong Lembah, Kuala Kangsar; Kampung Kota Lama Kanan and Kampung Bota Kanan, Parit.

Rumah Perak uses fully natural building materials found around the village. The basic structure is hardwood, the roof is made of thatch leaves, and the slate walls are made of woven reeds or rattan.

The description shows that the Kutai house is for ordinary people compared to more modern and sophisticated designs, such as the pyramid and cross-gable roofs. This

Figure 17 Rumah Perak

house has a high roof, but there is little uniqueness in terms of the openness of this house. It does not have good and sufficient ventilation. The house's walls reach the rooftop level without being crossed by filigree. This is because the walls themselves are porous, so there is no need for filigree decorations and the humility and modesty of the inhabitants of this house are sometimes confused with low self-esteem.

List of records of Stingless bees found in Perak state[11]. (Rasmussen 2008).

Record of a synonym to **Heterotrigona erythrogastra** (Cameron, 1902)

Syn. *Trigona luteiventris* Friese 1909("1908"): 354, 358, fig. 15-5: Lectotype (ZMHB, worker): here designated, "Malakka / Perak / 1912", "Trigona / luteiventris / 1904 Friese det. Fr / n Fr.", "Type" (red label), "Coll. /Friese"; paralectotype (ZMHB (1)); possible additional types (AMNH) (taxonomy); **Type locality**: MALAYSIA/PHILIPPINES "**Perak** (Malaka)"; "Palawan" (workers);

Record of a synonym to **Homotrigona fimbriata** (Smith, 1857)

Syn. *Trigona versicolor* Friese 1909("1908"): 358, 359, fig. 15-1: Lectotype (ZMHB, worker): here designated, "Tandjong / SO.-Borneo / coll. Speyer", "Trigona / versicolor / 1904 Friese det. / n. Fr.", "Type" (red label) (Comparative note, taxonomy); **Type locality**: MALAYSIA "Malaka (**Perak**); "SO-Borneo (Tandjong)" (2 workers);

Record of the previous combination for **Tetragonula basimaculata** (Bingham 1903)

Melipona basimaculata Bingham 1903: vii: Holotype (BMNH 17b.1127, male) (taxonomy); Type locality: MALAYSIA "Biserat, Jalor; Telôm, **Perak**-Pahang boundary, 4,000 feet" (male); Melipona basimaculata Bingham 1905: 60, plate A11 (citation).

Record of **Tetragonula reepeni** (Friese, 1918)

Trigona reepeni Friese 1918: 519-520: Lectotype (ZMHB, worker): here designated, "Malacca / Taip. Hills /2.1912 / Butt.-Reep.", "Trigona / reepeni / 1914 Friese det. / Fr.", "Type" (red label), "Coll. / Friese"; paralectotypes (DEI (1), USNM (1), ZMHB (7)) (taxonomy); **Type locality**: MALAYSIA "Taiping Hills (**Upp. Perak**) auf Malakka, im February 1912" (several workers);

Figure 18 Tetragonula basimaculata (Bingham 1903)

Record of previous combination **Tetrigona peninsularis** (Cockerell, 1927)

Trigona apicalis peninsularis subsp.n. Cockerell 1927: 541: Holotype (BMNH 17b.1099) (taxonomy, variation); **Type locality**: MALAYSIA "**Perak**, F.M.S., Batang Padang, Jor Camp, 1800 ft., June 4, 1923(Pendlebury)"; CAMBODIA, Patalung; MALAYSIA, Kuala Lumpur, Gombak valley;

Record of **Tetragonula melina** (Gribodo, 1893)

Trigona melina Gribodo 1893: 262-263, 264: Syntypes (MSNG, three workers; unknown depository, two workers): Two syntypes are labelled "Bandjarmas" (?=Bandjarmasin, near Liangtelan), while the third syntype carry no locality label (F. Penati, pers. com.) (taxonomy); **Type locality**: MALAYSIA "Liangtelan (Borneo)" (2 workers); "**Perak** (Malacca)" (3 workers);

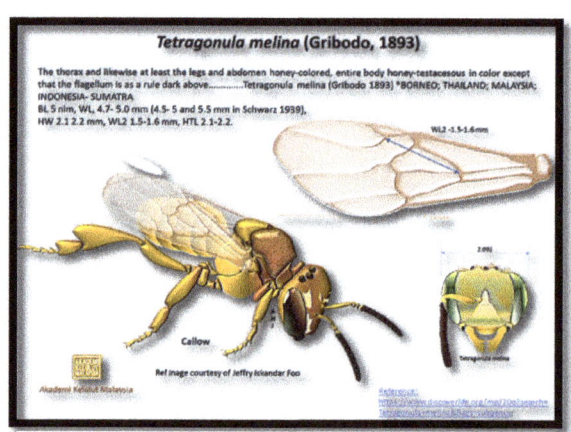

Figure 19 Tetragonula melina (Gribodo, 1893)

---

[11] This list provides local beekeepers to have a choice of species that may be found in their region and may prove suitable for culturing within their structures. It contains old names and synonyms (Rasmussen 2008) that may be familiar to the older generation of beekeepers.

2. Selangor House

This house is designed with a long roof in the state of Selangor. It contains a combination of a large and prominent platform and portico. All long roof designs are extended backwards when enlarged according to the needs of the house's current owner. This house does not have a terraced roof like the Negeri Sembilan house but has a large, steep roof in the 'mother' house. As is customary in this design, this large roof (a spread sail) aims to trap wind so that the roof space can be aired to lower the temperature inside.

Figure 20 In Selangor and the Federal Territory, the long-roof houses were influenced by the architecture of the long-roof houses in Melaka and Negeri Sembilan.

The bigger the 'mother' house, the bigger the roof. The exterior design of this house has a simple engraving on the top of the window. The window has a short rail that is located above the lower bar of the window. The window of this house is located slightly lower than the floor level. The 'mother' house or *Rumah Ibu* is elevated compared to the kitchen, pavilion, household, and porch. The height of the roof structure makes this house look spacious and relieved when inside. It also aims to make the airflow process fluent.

List of records of Stingless bees found in Selangor state (Rasmussen 2008).

Record of the previous combination for ***Pariotrigona pendleburyi*** (Schwarz, 1939)
   Trigona (Hypotrigona) pendleburyi variety ***klossi*** Schwarz 1939a: 85, 94, 132*: Holotype (BMNH, no number) (distribution, key to species, taxonomy); **Type locality**: MALAYSIA "MALAYA. State of **Selangor**: Bukit Kutu, 200 feet, Sept. 21, 1932, (H. M. Pendlebury)" (workers, holotype); "State of **Pahang**: Kuala Teku, 500 feet, Dec. 8, 1921 (H. M. Pendlebury)" (workers); "N. BORNEO. -Bettotan, nr. Sandakan, (C. B. Kloss and H. M. Pendlebury)" (workers, July 28, 1927, and Aug. 15, 1927);

Tetragonula minor (Sakagami, 1978)
   Trigona (Tetragonula) minor Sakagami 1978: 166-194, 213-214*, 224, 226, 227, 236, 237, 238: Holotype (SEHU); 32 paratypes (SEHU) (taxonomy); **Type locality**: MALAYSIA "Kuala Lumpur-2" (33 workers);

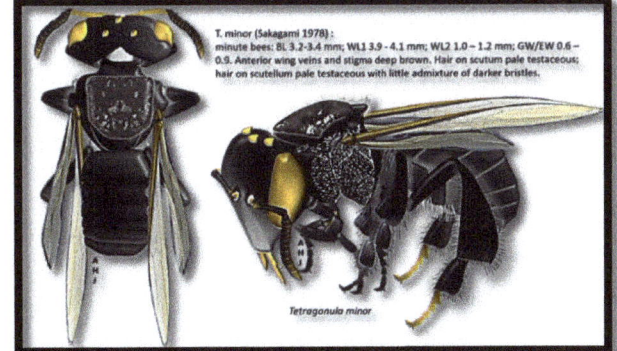

Figure 21 Tetragonula minor (Sakagami, 1978)

3. Melaka House

This house has a concise design and contains a basic function of space, showing the simplicity of the owner. This house design contains a porch, 'mother' house, and a kitchen house. In principle, this house is like the Kutai house of Perak. The difference is that there is a porch in the front of this house. The porch is a place to rest, breathe, serve guests, and review activities in front of the house.

Figure 23 In Melaka, the long roof house, also known as Rumah Serambi Melaka, is divided into a house of twelve pillars and a house of sixteen pillars.

Figure 22 Rumah Melaka

The entire wall of the house is made of wood, except for the part of the kitchen, which has a slate wall. The roof of this house is opened slightly to allow air to enter the house. The house hose is also on the wall, unlike the silver Kutai house. This house has a flower-patterned display and a simple openwork sign. The wooden bars on the porch also have concise carvings under the context of the simplicity of this Rumah Ibu and are more comfortable.

List of records of Stingless bees found in Malacca state (Rasmussen 2008).

Record of *Lepidotrigona latipes* (Friese, 1900):

Trigona latipes Friese 1900: 384: Holotype (ZMHB, worker): examined, "India / Singapore / 1890", "Trigona / latipes / 1909 Friese det. / Fr.", "Type" (red label), "Coll. / Friese". The nitidiventris species group (taxonomy); **Type locality**: SINGAPORE "**Malacca** (Singapore)" (1 worker);

Record of *Lepidotrigona nitidiventris* (Smith, 1857)

Trigona nitidiventris Smith 1857: 50-51: Type (OUMNH (=Wilson Saunders collection)). The nitidiventris species group (taxonomy); **Type locality:** MALAYSIA "**Malacca** (Mount Ophir)" (worker); Smith 1871: 395 (distribution);

Record of *Lepidotrigona ventralis* (Smith, 1857)

Trigona ventralis Smith 1857: 50: Type (BMNH 17b.1186); Mt. Ophir specimen is latebalteata. The ventralis species group (taxonomy); **Type locality:** MALAYSIA "Borneo (Sarawak)"; "**Malacca** (Mt. Ophir)"(worker);

Record of *Tetragonilla collina* (Smith, 1857)

Trigona collina Smith 1857: 51-52: Type (OUMNH (=Wilson Saunders collection)) (taxonomy); **Type locality**: MALAYSIA "**Malacca** (Mount Ophir)" (worker);

Record of T*etragonilla atripes* (Smith, 1857)

Trigona atripes Smith 1857: 50: Type (OUMNH (=Wilson Saunders collection)) (taxonomy); **Type locality**: MALAYSIA "**Malacca** (Mt. Ophir)" (worker);

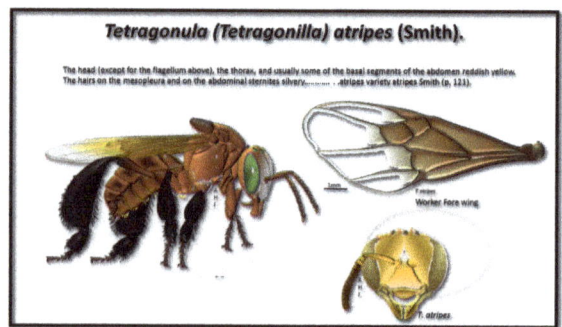

Figure 24 Tetragonilla atripes (Smith, 1857)

4. Penang Island House

Figure 25 Rumah Serambi Gajah Menyusu – Most of them are found in Pulau Pinang.

On Penang Island or *Pulau Pinang*, there are several types of traditional Malay houses with different shapes, namely the Long Bumbung House, Porch House and Nursing Elephant Porch. One of the traditional Malay houses, the Nursing Elephant Porch House, can be recognized by the shape of the roof where the 'mother' roof looks higher and lower as if an elephant is breastfeeding her child.

This type of house uses a long, horizontal type of roof, and in front of the edge of the wall, an upright wall is installed, called a *tebar layar*, when the division of space also consists of the 'mother' house and the kitchen. The space division consists of a household room in front of the outer foyer, the living room, the inner foyer, the 'suckling elephant' or *Gajah Menyusu* room and the kitchen. The building materials for this house consist of cengal and meranti wood (resinous woods of the Dipterocarpaceae family), while the roof is made of nipa thatch.

A significant stingless bee species associated with bee fauna in Penang is *Tetragonula penangensis* (Cockerell, 1919) [Typ. loc.: Malaysia "Penang Island (Baker 9075)"] (Rasmussen, 2008) = The epithet refers to Penang and the species distribution is Penang Island but quite uncommon and rarely seen elsewhere.

List of records of Stingless bees found in Penang state (Rasmussen 2008).

Record of a synonym to ***Tetragonula fuscobalteata*** (Cameron, 1908)

Syn. *Trigona atomella* Cockerell 1919b: 243-244: Holotype (USNM 29467, worker): examined, "Island of / **Penang** / Baker", "Type No. / 29467 / U.S.N.M.", "Trigona / atomella / Ckll TYPE"); 2 paratypes (USNM) (comparative notes, taxonomy); **Type locality**: MALAYSIA "Island of **Penang** (Baker, 9585)" (worker);

Figure 26 Tetragonula fuscobalteata (Cameron, 1908)

Record of ***Tetragonula penangensis*** (Cockerell, 1919)

Trigona penangensis Cockerell 1919c: 78, 79: Holotype (AMNH) (comparative notes, key to species); **Type locality:** MALAYSIA "**Penang** Island (Baker 9075)" (worker);

Record of ***Tetragonula zucchii*** (Sakagami, 1978)

Trigona (Tetragonula) zucchii Sakagami 1978: 166-194, 196, 197, 198, 208-210*, 228, 229, 236, 237, 238, 247, plate 5: Holotype (SEHU) (nest, taxonomy); **Type locality:** MALAYSIA "Fraser's Hill-b, Malaya" (9 workers, holotype, eight paratypes); "Cameron Highlands" (8 workers); "Gombak" (1 worker); "**Penang**" (1 worker); "Templer Park" (17 workers);

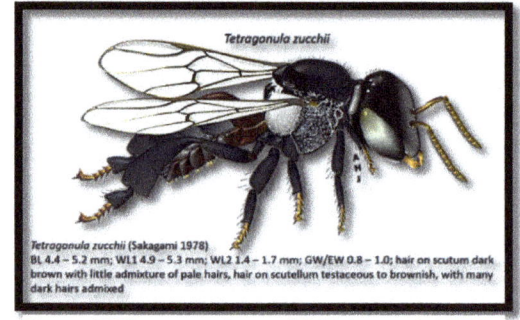

Figure 27 Tetragonula zucchii (Sakagami, 1978)

Record of ***Heterotrigona bakeri*** (Cockerell, 1919)

Trigona bakeri Cockerell 1919c: 78, 79: Holotype (USNM 29468, worker): examined, "Island of / Penang / Baker", "9069", "Type No. / 29468" (head glued to label), "Trigona / bakeri / Ckll TYPE" (comparative notes, key to species); **Type locality**: MALAYSIA "**Penang** Island (Baker 9068)" (worker);

Record of a synonym to ***Lepidotrigona terminata*** (Smith, 1878)

Syn. *Trigona fulvomarginata* Cockerell 1919c: 78: Holotype (BMNH 17b.1083) (key to species); **Type locality**:
MALAYSIA "**Penang**" (worker);

Record of ***Tetragonula testaceitarsis*** (Cameron, 1901)

Trigona testaceitarsis Cameron 1901: 36: Type (BMNH 17b.1121) (taxonomy); **Type locality**: MALAYSIA "Patani, Malay Peninsula" (worker);

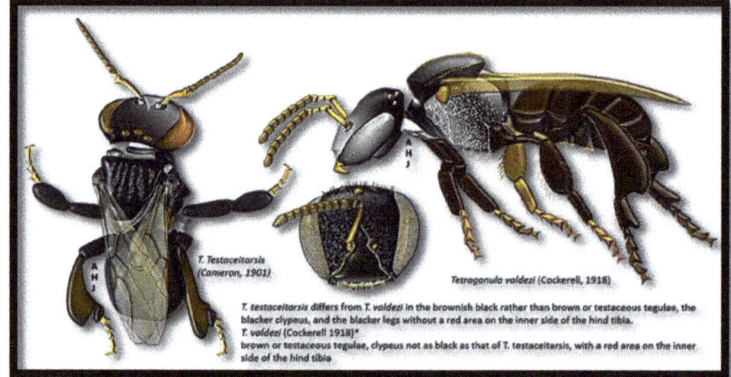

Figure 28 Tetragonula testaceitarsis (Cameron, 1901)

5. Kedah House

In Kedah, traditional houses are much reduced because many have been torn down and replaced with modern houses. The old traditional houses are beautifully decorated with carvings. The construction of traditional houses illustrates the high quality of the arts of the ancient Malays. The surrounding conditions and culture influence the construction of their homes. This influence is highlighted in the carvings found in their homes, such as plants, life and objects.

Figure 29 In Kedah and Perlis, Malay houses with long trunks are known as Rumah Serambi Kedah @ sometimes called longhouses only.

The house of the Malays of Kedah used to have the *Serambi* (porch), *Rumah Ibu* (mother or main house), *Rumah Tengah* (middle house) and *Dapur* (kitchen). The usually found is an awning roof and a long roof. If we look at the remains of this heritage, we can say that the Malays placed so much importance on the beauty and uniqueness of their homes. They are incredibly careful in building houses in terms of construction and site selection so that they will live in peace and security for all time.

A related species in the locality: **kusutkana**, *Trigona* Dover 1929 [Typ. Loc.: Malaysia "Kedah Peak."]# (Rasmussen, 2008) kusutkana refers to the Riau Malay phrase "*kusut kana*", meaning "tussled or ruffled because of implies hair being tussled or ruffled after an attack by a group of these bees.

Record of a synonym to ***Homotrigona aliceae (Cockerell, 1929)***
  Syn. *Trigona kusutkana* Dover 1929: 62-64: Holotype (not in ZRC); paratype (not in ZRC) (taxonomy, variation); **Type locality**: MALAYSIA "**Kedah** Peak, 3200 ft., December 1915" (holotype worker, paratype worker) (Rasmussen 2008).

6. Negri Sembilan House

Negeri Sembilan's traditional houses exhibit a long roof design characteristic of Minangkabau. The lower part of the roof and the house's walls are covered with large screen panels, thus providing a strong enough identity with its aesthetic value. At the same time, the screen's width is carved, thus allowing the flow of air into the roof space.

Loaded with uniqueness, traditional Negeri Sembilan houses are characterized by multi-storey roof designs covered with thatched nipa leaves. Usually, it has a front porch and a porch that shades the main staircase to enter the 'mother' house through the porch. The porch is decorated with crisscross patterned filigree carvings.

The 'motherhouse' is open and is connected to the kitchen through a door that leads to an interval that does not have walls and becomes a platform for part of the kitchen's functions. It acts as a chat room for women.

Figure 31 The long roof that is Minangkabau

Figure 31 Rumah Warisan Negri Sembilan

The porch and the 'mother' house have various levels. 'Mother' or main house has the widest and highest space. There is a room in the main house and a staircase that leads to the attic. Usually, the attic is used as a bedroom for single boys because the steepness of the stairs is not suitable for girls.

Many stingless bee populations were recorded in Pasoh Forest Reserve, Negri Sembilan (San & Osawa 2002, Appanah 1982). Among the uncommon ones are **Lisotrigona cacciae** (Nurse, 1907) syn. *Trigona scintillans* Cockerell 1920a: **Lophotrigona canifrons** (Smith, 1857) syn. *Trigona busara* Cockerell 1918: and **Sundatrigona moorei** (Schwarz, 1937) syn. *Trigona (Tetragona) matsumurai* Sakagami 1959.

7. Pahang House

Figure 33 Rumah Pisang Sesikat

Figure 33 Rumah Pahang

The Pahang Malay Traditional House is known as the *Rumah Serambi Pahang*. This type of house is disappearing. It usually has a long ridge when a screen is attached to the edge of the wall and carvings are on the edges. Its components consist of the *Rumah Ibu* (mother or main house) and the kitchen, and the two are separated by a space called *anjung*. It also contains certain spaces at the front, namely the foyer, followed by the 'mother' room, which contains the stage and hall. A child's armrest room or back porch is also in the back portion. The kitchen contains a banquette and a platform. These spaces are used for waiting, cooking and so on.

A record of a species in this locality is that of ***Pariotrigona pendleburyi* (Schwarz, 1939)**

> ***Trigona (Hypotrigona) pendleburyi*** Schwarz 1939a: 85, 86, 94, 130-132*: Holotype (BMNH, no number); paratypes (BMNH (5)) (distribution, key to species, taxonomy); **Type locality:** MALAYSIA "Cameron Highlands, Rhododendron Hill, 5000 feet (H. M. Pendlebury)" (workers, holotype, June 19, 1923, and June 22, 1923); MALAYA. -State of Pahang: Cameron Highlands, G. Tarbakar, 4500 feet, June 9, 1923 (H. M. Pendlebury); Cameron Highlands, Bukit Lendong, 5000 feet, May 21, 1931 (H. M. Pendlebury); Lubok Tamang, 3500 feet, March 5, 1924, and 4000 feet, June 10, 1923 (H. M. Pendlebury); Fraser's Hill, 4600 feet, Pine Tree Hill, Oct. 25, 1932 (H. M. Pendlebury) (Rasmussen 2008).

8. Johor Limas House

The Johor Limas house is built high in the same basic form. The way to identify a Johor Limas house is through the roof of the house, which has a long roof and is connected to four pyramids. A sharp wood is installed at the top of the roof, called *tunjuk langit* (point to the sky). This house has a front

Figure 36 Rumah Limas Johor (Gable and Hip Roof)   Figure 36 Johor: Rumah Potong Belanda @ Rumah Limas

room called the 'mother' house, and at the end is the pavilion. Usually, this house is made of Cengal, Keranji, Perak and Meranti wood.

Figure 34 Left Palembang Limas House Source: https://www.wikiwand.com/en/Malay_hous ;Right: Johor Limas House

A village-style landscape is built in the yard of this Palembang Limas house. It looks beautiful and understated. Planting verdant grass makes the garden area fresher, especially with the combination of the colour of the house's wood swept with wood dye.

9. Kelantan House

In Kelantan, there are two traditional Malay houses: *Rumah Bujang* (Bachelor House) and *Rumah Tiang Dua Belas* (Twelve Pillar House). The ridge is long, where the corner edges of the walls are fixed with screens. At the edge of the screen is a rafter's board or pedestal of the rafters, namely a pair of 'mother' handlers and a pair of porch beams.

Figure 37 Houses in Kelantan, there are two types of long-roofed houses which are Rumah Tiang Dua Belas and Rumah Bujang.

The original Tiang 12 house has three building units, namely the 'mother' house, which contains a lobby, a central house, and a kitchen house. The middle house and kitchen house are cubicles or insulated.

10. Perlis House

The Perlis Malay traditional house is a long-roof or 'Bumbung Panjang' consisting of the *Rumah Ibu* (mother or main house) and the kitchen house. There is a hall room connected with the 'mother' house. This hall in front of the kitchen and between the 'mother' house has both functions in extension. The building is usually made of wood or reed, roofed with thatch or palm leaves.

Figure 38 Rumah Perlis

Usually, this house is walled with reeds and *kelarai* (bark or *mengkuang,* i.e., screw pine leaves woven with slats or rattan), with plank floors. The upper room of the window has carvings on the roof, as well as on the rake and fascia boards and 'screen display'. Only the wealthy will use the upright, tongued, and prostrate board. Source[12]:

---

[12] http://v12gether.blogspot.com/2013/02/info-rahsia-rumah-melayu-lama.html
http://bicarasenivisual.blogspot.com/2014/09/seni-bina-rumah-melayu-tradisional.html

11. Terengganu House

Figure 39 Terengganu Malay House

Most of the Malay houses are built as Stage Houses *Rumah Panggung*. The main feature of the traditional Malay *'kampung'* house is that it is built on pillars or stilts. This elevation was to ward off wild animals and floods, deter thieves, and improve aerospace. The Terengganu house's prominent features are the rake board and the conspicuous finial known as 'tunjuk langit'. The main house incorporates an adjacent veranda for leisure.

Figure 40 In Terengganu, there are two types of long-roof houses, namely Rumah Tiang Dua Belas and Rumah Bujang.

12. Sarawak House

The Malays are part of the population living in Sarawak. The uniqueness that can be seen from the design of this house is that this house is built with high pillars several meters from the ground. This design was developed to provide artistic value, prevent the house from being easily flooded and reduce the threat of poisonous insects on the ground, such as centipedes, scorpions, etc.

The design of this house is still strongly influenced by Malay culture, with the overall design built using wood and tiled roofs. The wood has been carved with carvings or filigree to add more art to the decoration, especially on the top of the windows, stair railings ('dak-dak' in Sarawak Malay) or the lattice. Among the built structures that make up this house are skylights, belian roofs, latticework, and stairs.

Figure 42 A traditional Malay Sarawak house in Malaysia.
Source: https://www.wikiwand.com/en/Malay_house

Figure 42 Traditional Malay Sarawak house

List of records of Stingless bees found in Sarawak state (Rasmussen 2008).

Record of ***Tetragonula sarawakensis*** (Schwarz, 1937)
    Trigona sarawakensis Schwarz 1937: 283, 290, 313-315*, 316, 328, plate 2, 4: Holotype (BMNH 17b.1110); paratypes (AMNH, BMNH?) (Key to species, taxonomy); **Type locality**: MALAYSIA "**Sarawak**: Mt. Dulit, 4000 ft." (2 workers Oct. 18, 1932; 2 workers Oct. 19, 1932; 1 worker Oct. 22, 1932);

Record of ***Lepidotrigona latebalteata*** (Cameron, 1902)
    Trigona latebalteata Cameron 1902: 130-131: Syntypes (BMNH 17b.1084, 2 workers). The terminata species group (taxonomy); **Type locality**: MALAYSIA "Kuching, **Sarawak**" (worker);

Record of ***Heterotrigona erythrogastra*** (Cameron, 1902)
    Trigona erythrogastra Cameron 1902: 129-130: Holotype (BMNH 17b.1130); 1 paratype (BMNH?) (Comparative notes, taxonomy); **Type locality**: MALAYSIA "**Sarawak** (R. Shelford)" (worker);

Record of a synonym to ***Homotrigona fimbriata*** (Smith, 1857)
    Trigona flavistigma Cameron 1902: 130: Holotype (BMNH 17b.1097) (taxonomy); **Type locality**: MALAYSIA "Kuching, **Sarawak**" (male);

Record of ***Lepidotrigona latebalteata*** (Cameron, 1902)
    Trigona latebalteata Cameron 1902: 130-131: Syntypes (BMNH 17b.1084, 2 workers). The terminata species group (taxonomy); **Type locality**: MALAYSIA "Kuching, **Sarawak**" (worker);

Record of ***Lophotrigona canifrons*** (Smith, 1857)
    Trigona canifrons Smith 1857: 51: Syntype (OUMNH, one worker); additional putative type (BMNH 17b.1183) (taxonomy); **Type locality:** MALAYSIA "Borneo (**Sarawak**)" (worker);

Record of the previous combination for ***Odontotrigona haematoptera*** (Cockerell, 1919)
    Trigona haematoptera* variety *dulitae Schwarz 1937: 282, 283, 286, 291, 320-321*, 328, plate 2, 6: Holotype

(BMNH 17b.1125); paratypes (AMNH, BMNH (8)) (key to species, taxonomy); **Type locality**: MALAYSIA "**Sarawak**: Foot of Mt. Dulit, junction of rivers Tinjar and Lejok" (2 workers Aug. 2, 1932; 10 workers Aug 3, 1932); "Mt. Kalulong, 1800 ft." (1 worker);

Record of *Borneotrigona hobbyi* (Schwarz, 1937)
Trigona hobbyi Schwarz 1937: 283, 288, 298-300*, 328: Holotype (BMNH 17b.1118). The hobbyi species group (key to species, taxonomy); **Type locality**: MALAYSIA "**Sarawak**: Mt. Dulit, 4000 feet, moss forest, Oct. 18, 1932" (1 worker);

Figure 43 Borneotrigona hobbyi (Schwarz, 1937)

Record of *Sundatrigona moorei* (Schwarz, 1937)
Trigona moorei Schwarz 1937: 283, 292, 321-322*, 328, plate 3, 4: Holotype (BMNH 17b.1124); 1 paratype (AMNH) (key to species, taxonomy); **Type locality**: MALAYSIA "**Sarawak**: Mt. Dulit, R. Lejok, near sweat and water, Oct. 5, 1932" (2 workers);

Record of *Tetragonula fuscobalteata* (Cameron, 1908)

Trigona fusco-balteata Cameron 1908: 193, 194: Lectotype (BMNH 17b.1112): Designated Moure 1961 (key to species, taxonomy); **Type locality**: MALAYSIA "Medang, **Sarawak** (Hewitt)" (worker); Schwarz 1937: 290, 310-311*, 316, 326 (key to species, taxonomy);

Figure 44 Tetragonula fuscobalteata (Cameron, 1908)

Syn. *Trigona pallidistigma* Cameron 1908: 193, 195: Syntypes (BMNH 17b.1133, 2? workers) (key to species, taxonomy); **Type locality**: MALAYSIA "**Sarawak**, Borneo (R. Shelford)" (worker);

Record of *Tetrigona apicalis* (Smith, 1857)
Trigona apicalis Smith 1857: 51: Holotype (OUMNH); additional putative type (BMNH 17b.1188) (taxonomy); **Type locality**: MALAYSIA "Borneo (**Sarawak**)" (worker);

Types and Variants of Malay Houses

Figure 46 A large Malay Hall originated from the Fold Kajang style and continued with the Limas roof style, Pavilion Riau, and Taman Mini Indonesia theme park. This style of structure is often used in the architecture of the palaces of Malay kings, and royal buildings. Source: https://www.wikiwand.com/en/Malay_house

Figure 46 Rumah Lancang or Rumah Lontik with a curved roof and a bot-like structure. A traditional Riau Malay house, this is the theme park of the Riau Pavilion Taman Mini Indonesia. Source: https://www.wikiwand.com/en/Malay_house

In Sumatra, traditional rigid houses are designed to ward off dangerous wild animals, such as snakes and tigers. While in an area located close to the major rivers of Sumatra and Borneo, the poles help to raise houses high above the flood level. In parts of Sabah, the number of dowry buffalo may also depend on the number of poles in the bride's family home.

- **Rumah Limas** is mostly religious in Palembang, Riau, Johor, Melaka, Pahang, Terengganu, and Selangor.
- **Rumah Lipat Kajang or Rumah Kejang Lako** - Mostly religious and are found in Jambi and Riau.
- **Rumah Melaka** – Most of them are located in Johor and Melaka.
- **Rumah Lancang or Rumah Lontik** – Most of them are found in Riau, Kampar Regency.
- **Rumah Belah Bubung** - Mostly religious found in the Riau Archipelago.
- **Rumah Kutai** – Mostly found in Perak and northern Selangor, based on Kutai architecture.
- **Rumah Perabung Lima** – Mostly found in Kelantan and Terengganu.
- **Rumah Gajah Menyusu** – Most of them are found in Pulau Pinang.
- **Rumah Tiang Dua Belas** – Mostly found in Kelantan, Terengganu and Pattani.
- **Rumah Bumbung Panjang** – Mostly found in Kedah, Perlis, Perak, Selangor, Johor and Pahang.
- **Rumah Air** – Mostly found within Brunei and Labuan.
- **Rumah Berbumbung Lima** – Mostly found in Bengkulu.

Figure 49 Left: Figure 39 Rumah Panggung; Right: Rumah Bujang Berserambi redrawn from Source: https://www.cgtrader.com/3d-models/architectural/other/rumah-bujang-berserambi-selasar

Figure 48 Traditional Bruneian Malay houses on stilts in Kampong Ayer, the traditional riverine settlement in Brunei. Source: https://www.wikiwand.com/en/Malay_house

Figure 48 Malay house in Sungai-liat, Bangka Island. Source: https://www.wikiwand.com/en/Malay_house

Figure 51 Rumah Belah Bubung - Most religions are found in the Riau Archipelago.

Figure 51 Rumah Kejang Lako in Rantau Panjang, Jambi

Rumah Panjang Kadazan Sabah or Sabah Kadazan Longhouse

This round-ended longhouse structure resembles the Sema Naga Houses in Nagaland and the Shan Tribal houses in Northern Myanmar. The difference in Sabah is they are built by the community for the dwelling of multiple families.

Traditional Lotud house in Sabah.

Figure 52 Rumah Panjang Kadazan Sabah or Sabah Kadazan Longhouse

People of the Lotud tribe in Tuaran, Sabah, build houses on stilts; some have Monitor or Clerestory windows in the attic. This way is specially built among communities living near the main Tuaran River.

Wooden A-Frame Tree House Bee Shed on stilts to alleviate above flood levels. A-frames below use Rect. MS hollow sections. A platform on stone pedestals.

Figure 54 Lotud house in the Heritage Village of Kota Kinabalu   *Figure 53 Inspired by the traditional Lotud tribal house.*

The rehabilitation of a traditional Lotud house in the Heritage Village of Kota Kinabalu, Sabah. Source : https://commons.wikimedia.org/wiki/ Category:Traditional_houses_in_Sabah#/media/ File:Lotud_House.JPG

A model of a longhouse is displayed in the Kinabalu Park Museum. The line of gongs in the foreground function as doorbells, and most longhouses have one so visitors can ring before entering. Each gong produces a different tone, and sticks are provided to find out for yourself. You have to bang on the

bulge, the bulbous part at the centre of the gong, and sticks are provided for the purpose. There is a polite notice requesting you not to bang too hard. (Kota Kinabalu, East Malaysia, Nov. 2013)   Source:

Figure 55 Model of a traditional Sabah House

https://commons.wikimedia.org/wiki/Category:Traditional_houses_in_Sabah#/media/File:Model_of_a_traditional_Sabah_house_(11967833764).jpg

Rumah Panjang Murut Sabah or Sabah Murut Longhouse

Figure 56 Inspired by the Murut Longhouse of Sabah.

The Murut are an indigenous ethnic group comprising 29 sub-ethnic groups inhabiting the northern inland regions of Borneo. The Murut tribe of North Borneo have their community longhouse on ornately carved pedestals made from 'Belian' (Eusideroxylon zwageri) or Bornean Ironwood. The roof structure resembles sails, and the pediments are adorned with tribal motifs. Interestingly, The posts are kept and moved around to form longhouses for at least 100 years. The Murut is mainly in Sabah, Malaysia, Sarawak, Malaysia, Brunei, and Kalimantan, Indonesia.

List of records of Stingless bees found in Sabah state (Rasmussen 2008).

Record of a synonym to ***Geniotrigona lacteifasciata*** *(Cameron, 1902)*
   Trigona lacteifasciata Cameron 1902: 131: **Holotype** (BMNH 17b.1132). Probably a distinct taxon of t*horacica* from the island of Borneo (taxonomy); **Type locality**: MALAYSIA "**Borneo**" (unknown);

Record of a synonym to ***Heterotrigona erythrogastra*** (Cameron, 1902)
   Syn. *Trigona sandacana* Cockerell 1919b: 242-243: Holotype (BMNH 17b.1129) (comparative notes, taxonomy); **Type locality**: MALAYSIA "**Sandakan, Borneo** (Baker, 9593)" (worker);

Record of a synonym to ***Geniotrigona thoracica*** (Smith, 1857)
   Syn. *Trigona ambusta* Cockerell 1918: 387: Holotype (BMNH 17b.1131) (distribution, key to species); Type Locality: MALAYSIA/SINGAPORE "**Sandakan** and Singapore" (unknown);

Record of a synonym to ***Heterotrigona itama*** (Cockerell, 1918)
   Syn. *Trigona breviceps* Cockerell 1919b: 244: Holotype (BMNH 17b.1123, worker): examined, "Type" (red border), "Sandakan / Borneo / Baker", "Trigona / breviceps / Ckll TYPE", "B.M. TYPE / HYM. / 17B.1123", "9591" (comparative notes, taxonomy); **Type locality**: MALAYSIA "**Sandakan, Borneo** (Baker, 9591)" (worker); Cockerell 1920b: 228 (distribution);

Record of a synonym to ***Homotrigona anamitica*** (Friese, 1909)
   Syn. *Trigona melanotricha* Cockerell 1918: 386, 387: Holotype (BMNH 17b.1095) (key to species, comparative notes, key to species, taxonomy); **Type locality**: MALAYSIA "**Sandakan, Borneo** (Baker 9222)" (worker);

Record of ***Lepidotrigona trochanterica*** (Cockerell, 1920)
   Trigona trochanterica Cockerell 1920a: 115: Holotype (BMNH 17b.1102) (taxonomy); **Type locality**: MALAYSIA "**Sandakan, Borneo** (Baker)" (worker);

Record of a synonym to ***Lisotrigona cacciae*** (Nurse, 1907)
   Syn. *Trigona scintillans* Cockerell 1920a: 116: Holotype (BMNH 17b.1115) (taxonomy); **Type locality**: MALAYSIA "**Sandakan, Borneo** (Baker)" (worker);

Record of ***Odontotrigona haematoptera*** (Cockerell, 1919)
   Trigona hæmatoptera Cockerell 1919b: 243: Holotype (BMNH 17b.1126) (comparative notes, taxonomy); **Type locality**: MALAYSIA "**Sandakan, Borneo** (Baker, 9592)" (worker);

Record of ***Tetragonilla fuscibasis*** (Cockerell, 1920)
   Trigona fuscibasis Cockerell 1920a: 115-116: Holotype (BMNH 17b.1082) (taxonomy); **Type locality**: MALAYSIA "**Sandakan, Borneo** (Baker, 9964)" (worker);

Record of ***Tetragonilla rufibasalis*** (Cockerell, 1918)
   Trigona rufibasalis Cockerell 1918: 387: Holotype (BMNH 17b.1111) (key to species, comparative notes, taxonomy); **Type locality**: MALAYSIA "**Sandakan, Borneo** (Baker 9225)" (worker);

Adaptation of Vernacular structures for Meliponary designs.

On the one hand, while these structures provide improved thermal comfort, the downside is that some beekeepers do not favour them. Firstly, the materials are not readily available in these times and age. Secondly, thatch roofing tends to harbour spiders and geckos, fierce bee predators. Thirdly, lumber needs drying before use to minimise warping or distortion and to eradicate fungus spores. Finally, untreated wood and bamboo may aggravate borer insects and interfere with the bee pheromones if chemically treated.

Figure 57 Vernacular structures in North Peninsula Malaysia

Other Ethnic Houses around Malaysia

Figure 58 A typical Orang Asli stilt house in Ulu Kinta, Perak.

Orang Asli (lit. "first people", "native people", "original people", "aborigines' people", or "aboriginal people" in Malay) are a heterogeneous indigenous population forming a national minority in Malaysia. They are the oldest inhabitants of Peninsular Malaysia. Source[13]:

Nabawan, Sabah: The framework of a traditional native house. The villagers undertook the construction. Source[14]. The image is credited with "Photo by CEphoto, Uwe Aranas."

Figure 59 The framework of a traditional native house.

---

[13] https://en.wikipedia.org/wiki/Orang_Asli
[14] https://commons.wikimedia.org/wiki/Category:Traditional_houses_in_Sabah#/media/File:Nabawan_Sabah_Construction-of-a-traditional-house-01.jpg

The Melanau Traditional House

The Melanau ethnic group comprises 6% of the population in Sarawak. A few interesting facts: sago is a traditional food for these people. Regarding guidance, the Melanau people have a longhouse as their traditional house. The longhouse is built so that it is 40 feet above the ground. What's interesting is that this house was built in such a way as to ward off enemy attacks, especially pirates from the sea.

Figure 60 Melanau Traditional House in the Sarawak Cultural Village in Kucing, Sarawak.

Malay Culture, Architecture and Literature

Translated excerpts from https://ms.wikipedia.org/wiki/Kebudayaan_Malaysia

Malaysia is a multi-ethnic, multicultural, and multilingual society, and many ethnic groups in Malaysia maintain distinct cultural identities. Malaysian society has been described as "Asia in miniature". The original culture of the area originated from the indigenous tribes, along with the Malays who moved there in ancient times. A great influence exists from Chinese and Indian culture since trade with those countries began in the area. Other cultures that greatly influenced Malaysia include Persian, Arabic, and English. The government structure and the balance of racial power caused by the idea of the social contract have caused little incentive for the cultural assimilation of ethnic minorities in Malaya and Malaysia. The government has historically made little distinction between "Malay culture" and "Malaysian culture".

The Malays, who account for more than half of Malaysia's population, play a dominant role politically and are included in the group identified as *bumiputra*. Their mother tongue, Malay, is the national language of the country. By the definition of the Malaysian constitution, all Malays are Muslims. *Orang Asal*[15], the earliest inhabitants of Malaya, made up only 0.5 % of the total population in Malaysia in 2000 but represented the majority in East Malaysia. In Sarawak, most non-Muslim groups are classified as Dayaks, making up about 40 % of the state's population. Many tribes have converted to Christianity. 140,000 *Orang Asli* Indigenous people comprised several different ethnic communities in Peninsular Malaysia.

---

[15] The Orang Asal are the indigenous people of Malaysia. The term is Malay for "Original People", used to refer to the aboriginals of Sabah, Sarawak, and Peninsular Malaysia. These groups are given the Bumiputra status in Malaysia.

Architecture in Malaysia is a combination of various styles, from Islamic and Chinese styles to those brought by European colonists. Malay architecture has changed due to these influences. The houses in the north are the same as in Thailand, while in the south are the same as in Java. New materials, such as glasses and nails, were brought by Europeans, changing the architecture. Houses were built for tropical conditions, raised on stilts with high roofs and large windows, allowing air to flow through the house and cool it. Wood has been the main building material in Malaysian history; it was used for everything from humble villages to royal palaces. In Negeri Sembilan, traditional houses are completely free of nails. Besides wood, other common materials, such as bamboo and leaves, are also used. Istana Kenangan in Kuala Kangsar was built in 1926, and it is the only Malay palace with bamboo walls. The *Orang Asal,* or indigenous people of East Malaysia, lived in long houses and water villages. Longhouses are raised on stilts and can house 20 to 100 families. Water villages were also built on stilts, with houses connected by planks and most transport by boat.

Chinese architecture can be divided into two types: traditional and Baba Nyonya. Baba Nyonya's household is made of colourful tiles and has a large inner courtyard. Indian architecture came with the Indian people of Malaysia, reflecting the southern Indian architecture that most originated from. Some Sikh architecture was also imported. Melaka, which is a traditional trading centre, has a variety of building styles. Large wooden structures such as Istana Sultan Mansur Shah (Figure 5) existed earlier. Chinese influence can be seen in brightly decorated temples and multi-story shophouses. The most remaining Portuguese structure in Malacca is the "A'Famosa" fort. Other colonial buildings include the Dutch Stadthuys, the brick buildings of the Dutch Colonial town, and English-built buildings such as the Hall of Remembrance, which combines Baroque and Islamic architecture.

The shape and size of houses vary from state to state. Common elements in Peninsular Malaysia include roofs, balconies, and high ceilings, which are raised on stilts for ventilation. Woodwork at home is often carved in situ.

Figure 61 Kg TambulianTraditional houses in Sabah

Tambulian, Kota Belud, Sabah: Heritage house project. The traditional houses are located on the green, opposite SK Tambulian. Source[16]:

---

[16] https://commons.wikimedia.org/wiki/File:Tambulian_KotaBelud_Sabah_Traditional-houses-01.jpg

Literature: A strong oral tradition that has existed since before the advent of writing has been transformed into what is continuing in Malaysia today. Indian epics heavily influenced these early works. Oral literature, such as folklore, flourished even as post-printed works appeared.

The first Malay literature was in the Arabic script. The earliest Malay writing is on the Kedukan Bukit Inscribed Stone, found by M. Battenburg. The earliest Malay writing is on the Terengganu Inscribed Stone, made in 1303. One of the more famous Malay works is "Sulalatus Salatin", also known as Sejarah Melayu (Malay History).

Figure 62 Melaka Literature Museum, Melaka City, Melaka, Malaysia. Source: https://commons.wikimedia.org/wiki/File:Melaka_Literature_Museum.jpg

Bangka-Belitung Islands

Figure 63 Rumah Adat Beliting: Rumah Rakit / Rumah Gede

Rumah Adat Traditional: Rumah Rakit / Rumah Gede

The physical form and function of the physical form of traditional house of Belitong, known as *Rumah Panggong*, is supported by wooden poles or stubs at best. All the building materials are made of wood up to the roof, called wooden shingles. The house is divided into three parts: a patio room, a living room or main formal space, a buffer space at the back terrace, like a living room for leisure, and an informal family room when formal issues occur at the Rumah Gede.

Rumah Brunei (Brunei House), also called Rumah Air, is typically built over coastal waters. Source[17]:

Figure 64 Rumah Brunei (Brunei House) in the Heritage Park of Sabah Museum

---

[17] https://commons.wikimedia.org/wiki/Category:Traditional_houses_in_Sabah#/media/File:KotaKinabalu_Sabah_Rumah-Brunei-02.jpg

Vernacular architecture and Insect Tourism

by Abu Hassan Jalil

In North Peninsula Malaysia, ornate wood carvings of Malay architecture are adopted in beehive designs. For example, there are hives in a Meliponary behind the Arts and culture centre in Langkawi Island, Malaysia. Being a tourist destination, this cultural centre includes the Meliponary and features

Figure 65 Ornately carved facade of bee box hives in Langkawi Island.

sales of stingless bee honey. The objective here is for some conscientious stingless bee keepers to have their hands in the 'The Bee Tourism' cookie jar. These are typical Malay house replicas incorporated into the beehive design. The box cover extension and the roof pediment facade have the carved-through motifs of a traditional Malay wood carving of the bakawali plant (*Epiphyllum oxypetalum*).

The main feature of the bakawali flower was transformed according to the creativity and versatility of craftsmen[18].

The origin of motifs in Malay woodcarving in the Malay Archipelago may date back to the pre-Islamic era when the Malays were practising Hinduism and Buddhism. The motifs are Kala Makara, Gunungan, Stupa, Garuda, and Naga (Farish and Khoo, 2003).

Selected Bee Box hives model Malay House with an ornate facade

Figure 66 Selected Bee Box hives model Malay House with ornate facade – compilation from anonymous beekeepers

---

[18] N. Shaffee & I. Said - Types of Floral Motifs and Patterns of Malay Woodcarving in Kelantan and Terengganu. Procedia - Social and Behavioural Sciences 105 (2013) 466 – 475

Figure 69 Cross Gable roof models of box hives. From left: Omjol of Aceh, Middle; Anonymous; Right: Zaidi Ibrahim of Terengganu.

This cross-Gabble roof evolves into the cross-hipped and hip and valley roofs below.

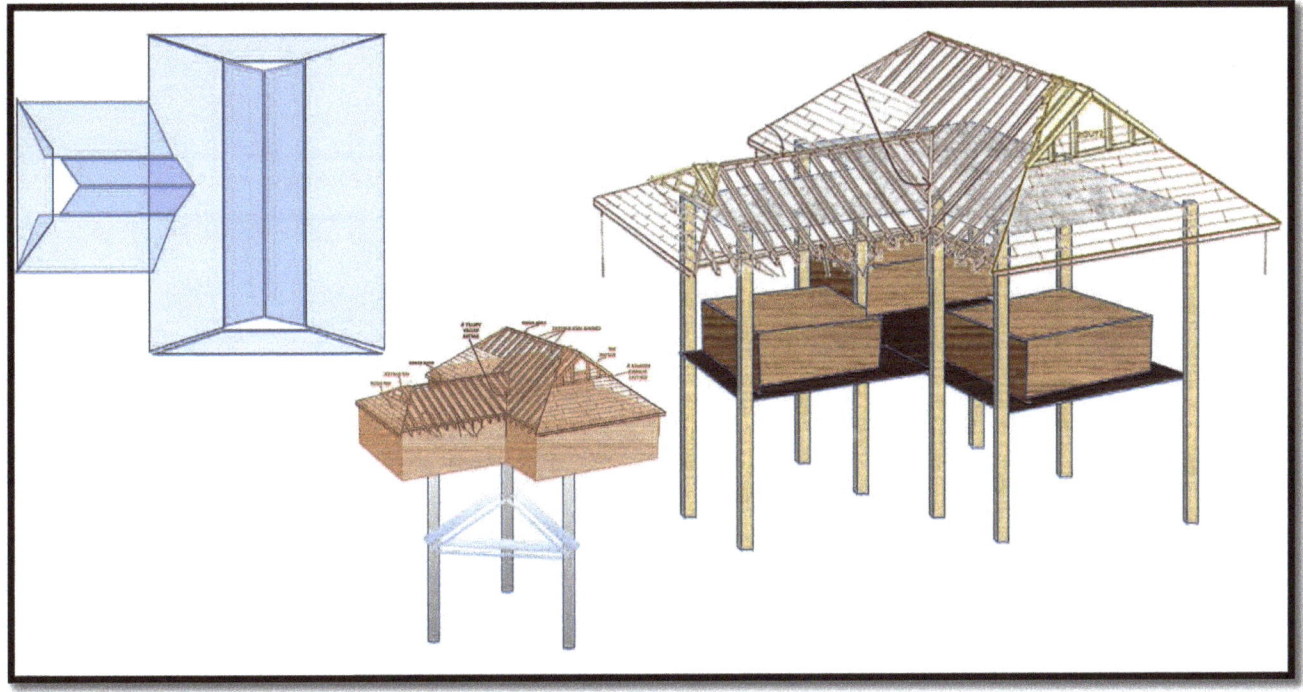

Figure 68 Cross-hipped roof applied on Bee rack and tri combo box hives with roof plan.

Figure 67 Traditional Malay wood carving used in box hive designs. Left: Terengganu Malay House; Middle and right: Lampung community hall design. Those on the right, although built-in Lampung, South Sumatra, the facade is of a motif derived from an ancient Javanese royal emblem and used in South Sumatra.

The traditional house turned Bee Gallery by AHJ.

Stingless beekeeper Izhar Johari of Bukit Wang, Jitra, Kedah, Malaysia, turned his house into a bee gallery. Initially, he had all his beehives in his backyard but eventually decided on a neat way to attract tourists.

Figure 71 Izhar Johari's house turned bee gallery which he said was inspired by a Terengganu house

Figure 70 Interior of his bee gallery with ornate woodcarvings

I visited Mr. Izhar Johari in Kedah while exploring the Northern state of Peninsular Malaysia. I was looking for the Dipterocarp-dependent species. I spent some time at his place because he was an avid woodcarver. Painstaking, he patiently completed this bee gallery over two years, frilled with elaborate floral carvings he designed.

Figure 72 Painstaking, he patiently completed this bee gallery over two years.

Box hive roofs

Izhar makes hilts for Malay daggers. He was also making unique roofs for his bee hive boxes. This image (figure 73) is one of his earlier hives where he was experimenting with roofs for the topping box of his hives. Rather basic initially, but as his beekeeping developed and his restless artistic soul emerged, he started looking at better and more elaborate designs to cater to the bee's comfort and the aesthetic value of his farm surroundings.

Figure 73 One of Izhar's earlier box hive roofs.

Figure 74 Malay stylized roof for a box hive

Although in the state of Kedah, Izhar says that his house was built in the design of a Terengganu traditional house. The carvings, though, are his creations of floral motifs. Those motifs are replicated on his carvings of the hilts and sheaths of the Malay daggers.

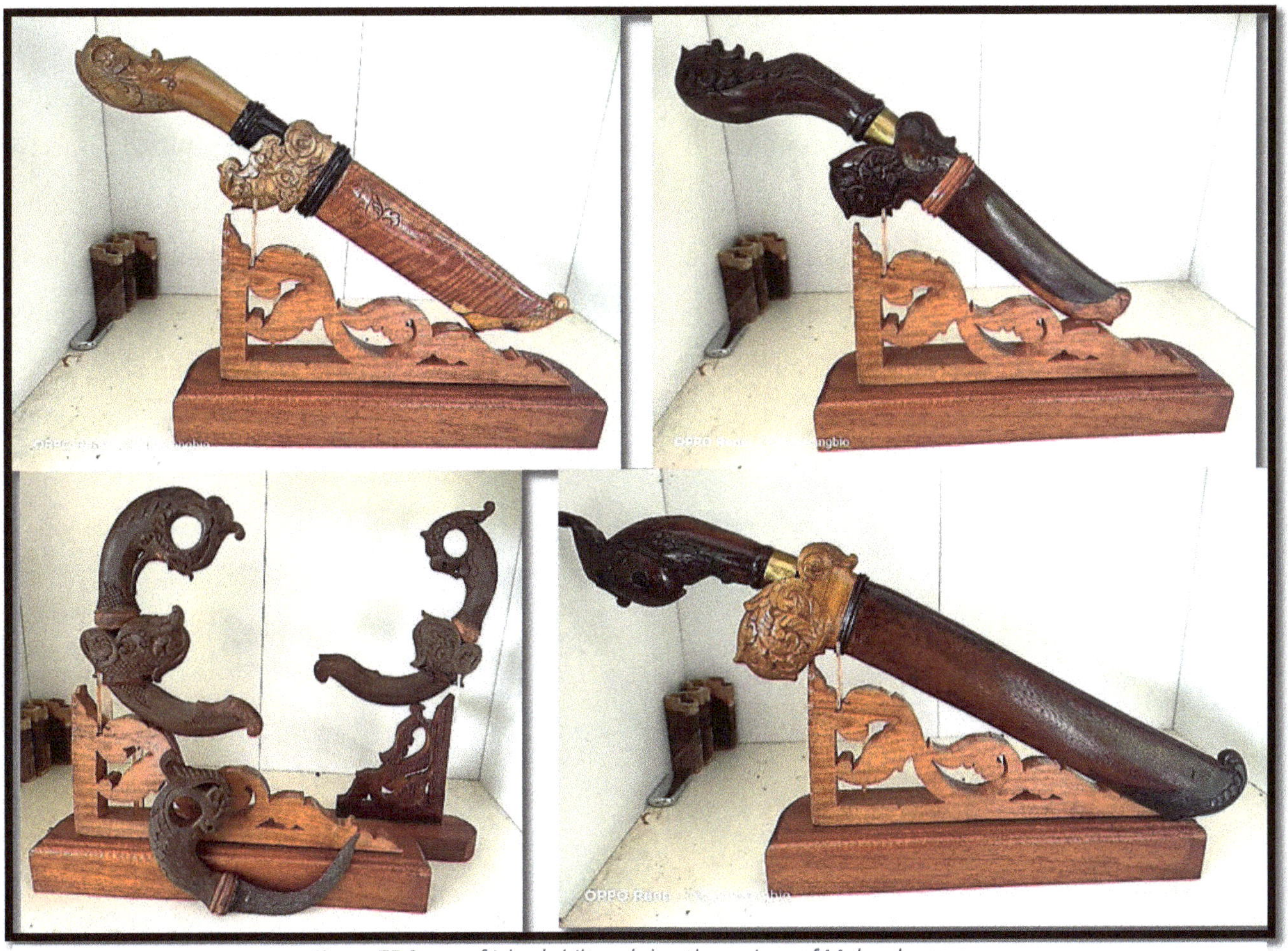

Figure 75 Some of Izhar's hilt and sheath carvings of Malay daggers.

Figure 76 Some of the touristy artefacts in the Gallery.

Min House Camp, Kubang Kerian, Kelantan by AHJ

In 2016, I visited Min House Camp in Kubang Kerian, Kelantan, owned by Wan Noriah Wan Ramli. The camp was named after her son Min (Abdul Muhaimin, who has Down Syndrome) and aimed at providing a "summer camp" for educating school children. They have a traditional Kelantanese house converted into a Bee gallery. It had an assortment of beekeeping paraphernalia, her kelulut products, and a collection of religious books. Interestingly, she keeps box hives facing glass louvred windows, and the bees can fly out to forage through them. Min House Camp has won an ASEAN tourism award before.

Figure 79 Collection of bee box hives in different styles for different species in the gallery. Bees forage through the louvred windows.

Figure 77 Outdoor hive.

Figure 78 Traditional Kelantanese structures at Min House Camp in Kubang Kerian, Kelantan.

Figure 80 Wan Noriah Wan Ramli, owner of Min House Camp in her Bee Gallery.

Figure 81 Bee hives are placed facing the glass louvres so that the bees can fly out to forage.

Big Bee Honey Gallery, Maran, Terengganu

by Mohd Razif Mamat

This modern-day Honey Gallery is in the suburbs of Terengganu. Although it lacks the vernacular ambience, it has the makings of a decently planned and excellent display of bottled honey and a few other products.

Figure 82 *Alhamdulillah, I had the opportunity to stop by to visit my friends at BBH Honey Gallery in Marang, Terengganu. I was impressed with the spirit shown by my friends. We wish them great success in the future. You can get honeybee honey and kelulut homey from him. Images and captions by Mohd Razif Mamat. May 2022.*

Vernacular architecture in the Malay Peninsula.

Figure 83 Vernacular architecture in the Malay Peninsula.

Chapter 3

Malay Nusantara traditional architecture

by Abu Hassan Jalil

This chapter examines the early origins and the evolution of Malay Nusantara traditional architecture.

Figure 85 Malay Nusantara traditional architecture redrawn from an anonymous website of public domain

Figure 84 Langkawi Island Malay house and a replica roofing on a log hive in the Arts and culture centre backyard in Langkawi.

Langkawi Island Malay house and a replica roofing on a log hive in Langkawi's Arts and Culture Centre's backyard. This roof design was incorporated into stingless bee box hives' roofing. A model roof built by En. Nasaruddin owns the Meliponary at the Arts and Culture Centre of Langkawi Island.

The original traditional Malay house developed into larger versions like the Bandung Institute of Technology (ITB)

While visiting a Meliponary in Bandung, we found this model among their hives. As it turns out, the maker was one of the students there in ITB.

Figure 86 West Wing of Bandung Institute of Technology (ITB) image from Wikimedia.

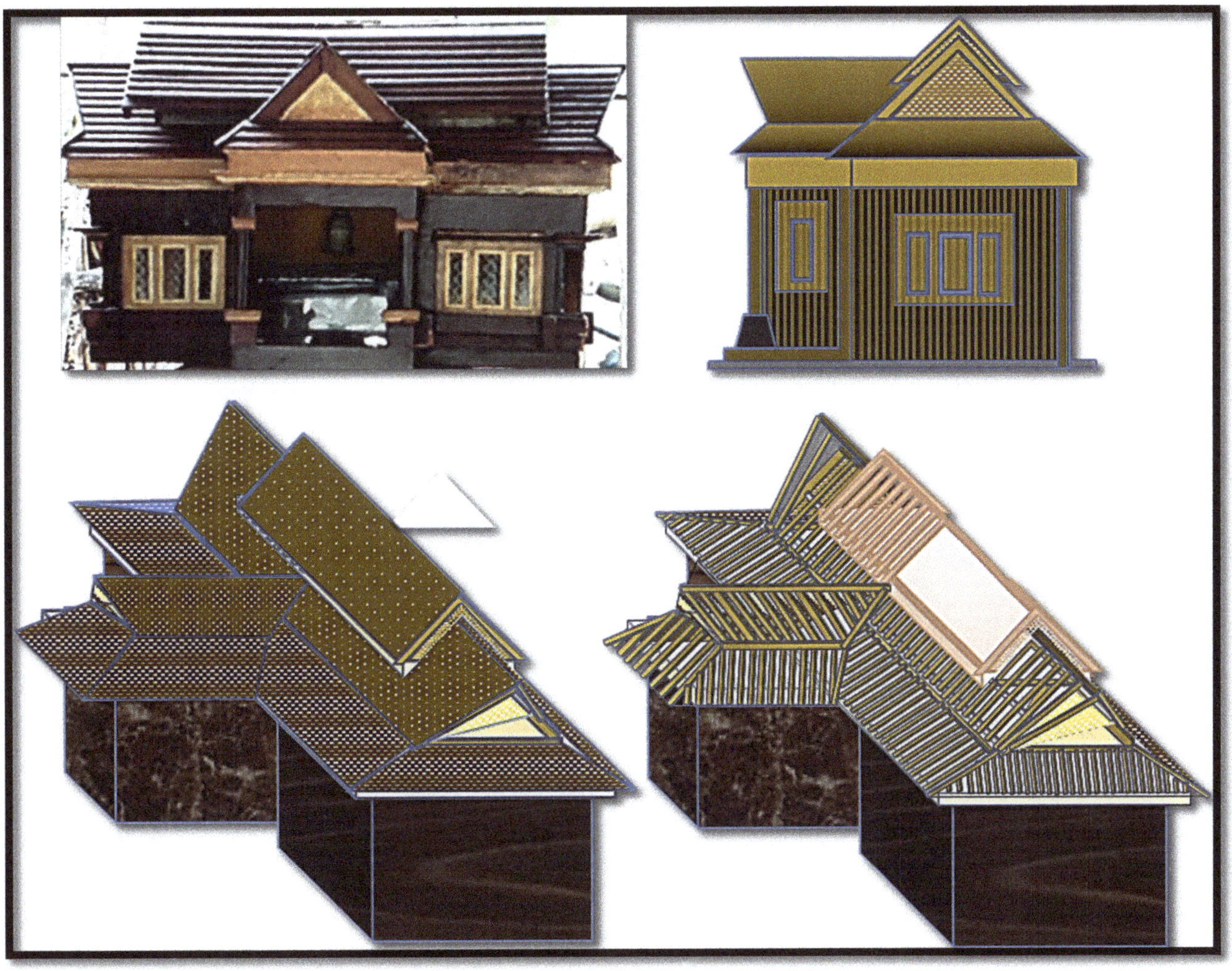

Figure 87 Illustration of the Box hive model in Meliponary belonging to Nzank (pronounced Enjang) in Bandung, Java

Figure 88 Exploded sequence of the Bandung model house with a Monitor roof

We also see it in a Jungle Trek Camp in Chiang Mai, Thailand, with a row of Kampung houses. (See Penang Island House – *Rumah Gajah Menyusu p.30 & p. 97*) One can see the power of symmetry in this architecture.

Figure 89 Figure 71 Jungle Trek Camp in Chiang Mai, Thailand, with a row of Kampung houses.

Malay Architecture Influence

It is time for me to fall back on Malay heritage and reminiscence on the elegance of Malay Architecture and its variants.

Figure 91 Example of Bees racks inspired by the Traditional tribal ethnic long house

Figure 90 Distribution of the Malay village house architecture in the Indo-Malaya ecozone.

Even the tribes Manabo & Maranao of Mindanao have the same basic gable and hip roof structure. Possibly, this is rooted in some archaic Polynesian influence.

Model making among beekeepers

A beekeeper named Cikgu Aman, a more apt name, Aman Londah, lives in Langkawi. Started from zero with only 6 logs of *Heterotrigona itama* at the start of 5 years ago. Now has become a successful business owner and conducts courses with several agencies.

Figure 92 Model Bee houses in Langkawi by Aman Londah

Another enthusiastic beekeeper, Luna Ahcmad Cahayani, an Indonesian living in Kuala Lumpur, has developed some simple Malay house model beehives for stingless bees.

Figure 93 Some Malay house model beehives for stingless bees made by Luna Ahcmad Cahayani

Malay Heritage Centre (MHC) in Singapore[19]

Istana Kampong Glam (Malay for "Kampong Glam Palace"; Jawi: ايستان كامڤوڠ ݢلم), also Istana Kampong Gelam, is a former Malay palace in Singapore. It is located near Masjid Sultan in Kampong Glam. The palace and compounds were refurbished into the Malay Heritage Centre in 2004.

Figure 94 Istana Kampong Glam right before restoration in August 2001

Sultan Hussein Shah of Johor built the original Istana Kampong Glam in 1819 on the land of about 23 hectares (57 acres) in Kampong Glam that had been given to him by the British East India Company. It is believed to have been a wooden structure east of Beach Road. When it was completed, it occupied an area twice the size of the present compound, which was reduced in 1824 to construct North Bridge Road.

List of records of Stingless bees found in Singapore (Rasmussen 2008)

Record of *Geniotrigona thoracica* (Smith, 1857)
Trigona thoracica Smith 1857: 50: Type (BMNH 17b.1181) (taxonomy); **Type locality: SINGAPORE** "Singapore" (worker);

Record of *Heterotrigona itama* (Cockerell, 1918)
Trigona itama Cockerell 1918: 387: Holotype (USNM 29471, worker): examined (distribution, key to species); **Type locality: SINGAPORE** "Singapore" (unknown);

Record of *Homotrigona fimbriata* (Smith, 1857)
Trigona fimbriata Smith 1857: 52: Type (BMNH 17b.1182) (taxonomy); **Type locality: SINGAPORE** "Singapore" (worker);

Record of *Lepidotrigona latipes* (Friese, 1900)
Trigona latipes Friese 1900: 384: Holotype (ZMHB, worker): examined, "India / Singapore / 1890", "Trigona / latipes / 1909 Friese det. / Fr.", "Type" (red label), "Coll. / Friese". The nitidiventris species group (taxonomy); **Type locality: SINGAPORE** "Malacca (Singapore)" (1 worker);

Record of a synonym to *Sundatrigona moorei* (Schwarz, 1937)
Syn. *Trigona (Tetragona) matsumurai* Sakagami 1959: 120-121: Holotype (SEHU, worker); 1 paratype (SEHU); **Type locality: SINGAPORE** "Singapore, Sept. 22, 1932, S. Matsumura leg." (2 workers);

---

[19] https://en.wikipedia.org/wiki/Malay_Heritage_Centre

Record of a synonym to *Tetragonula geissleri* (Cockerell, 1918)

Syn. *Trigona confusella* Cockerell 1919b: 242: Holotype (USNM 40248) (taxonomy, previously identified as **T. geissleri** (in Cockerell 1918)); **Type locality: SINGAPORE** "Singapore (Baker)" (worker); Schwarz 1937: 282, 290, 311-313* (key to species, taxonomy);

Figure 95 Tetragonula geissleri (Cockerell, 1918)

Record of *Tetragonula laeviceps* (Smith, 1857)

Trigona læviceps Smith 1857: 51: Holotype (OUMNH). Moure (1961) indicated that the BMNH type specimen (17b.1184) came from Mt. Ophir, which was not the type locality, and the specimen cannot be considered a true type. Baker (1993) located three specimens in OUMNH and labelled one as the holotype. Unfortunately, that specimen is identic to fuscobalteata, according to Baker (1993), but this issue will be resolved in a separate paper with C.D. Michener. Citations below for laeviceps are broad, as species limits are uncertain and may include valdezi and testaceitarsis. These taxa have previously been proposed as junior synonymous of *laeviceps* but are slightly larger (taxonomy); **Type locality: SINGAPORE** "Singapore" (worker);

Figure 96 Tetragonula valdezi (Cockerell, 1918)

Record of *Tetragonula valdezi* (Cockerell, 1918)
Trigona valdezi Cockerell 1918: 387: Holotype (USNM 40249, worker): examined; 3 paratypes (USNM) (Distribution, key to species, uncertain identity); **Type locality**: **SINGAPORE** "Singapore" (unknown);

Chapter 4
Sustainable Meliponiculture in Brunei
Meliponiculture in Tutong, Brunei by Mitasby Amit

Creative architecture in Box hives as a hobby of Mr. Tasby Amit

His Meliponine collection is also placed around his well landscaped house compound.

Tasby Amit, a pioneer in Meliponiculture in Brunei conducts classes in Tutong Village.

Figure 97 Meliponiculture in Tutong by Mitasby Amit in Brunei Darussalam

Figure 98 His Meliponary shed and Front view of Meliponine farm at Mitasby's house in Tutong, Brunei Darussalam.

Figure 99 a) b) c) Topping Box hive designs d) Peculiar Nest entrances collection

- For Vernacular Housing in Brunei, See *Figure 51 Traditional Bruneian Malay houses on stilts in Kampong Ayer, the traditional riverine settlement in Brunei.* Also, *Figure 64 Rumah Brunei (Brunei House) in the Heritage Park of Sabah Museum*

Urban Meliponiculture in Bandar Seri Begawan –

by: Mohd Fauzi Ja afar interviewed by Abu Hassan Jalil

In 2012, I began to make boxes for rearing stingless and managed to trap as many as three colony types of *Tetragonula sp.* in a fruit orchard area.

In 2013, with the help and instruction of master Tasby to collect 24 nests, *Heterotrigona itama*.

At first, I only bred at my house with one nest of *Geniotrigona thoracica*, one of *Heterotrigona itama* and one of *Tetrigona binghami*.

1 *Geniotrigona thoracica* nest results from cutting through a tree trunk to move the nest to the box hives at the landlord's request. It took 1-2 years for the colony to become stronger.

One nest, *Heterotrigona itama,* is also the result of transferring the hive to the hive box from a concrete block left behind with Tasby.

1 *Tetrigona binghami* encountered a nest on trees already fallen together with Tasby and tried to keep it at my residence locally, but the results were poor. *Tetrigona* species are not suitable for urban areas.

In 2015, almost all meliponine colonies in urban areas were moved to Ulu Tutong, near Tasik Merimbun.

At this time, I can collect 15 nests of *Tetrigona binghami,* two thoracic *Geniotrigona* nests, one nest of *Tetrigona melanoleuca,* one nest of *Lepidotrigona terminata,* 17 nests of *Tetragonula* sp, 8 *Tetragonula fuscobalteata* nest.

Harvesting methods initially once a month but now have slowed slightly to 4 months once following the harvesting of certain seasons. *Tetragonula* types are harvested by squeeze extraction of honeypots and other types using a portable suction pump as the honey pots are bigger.

In hunting new nests, we get nests from areas identified and planned for clearing for development and agriculture projects.

In terms of colony failures, it occurs only in the split colony trials ...

Splitting the last colony was successful when we started recognising males of species *G.thoracica*...

Honey production levels in 2014-2015 by 700-1000gm (24 nests) for B$ 15 (136 gm)

Production only in Jan-Aug and November to Dec

The above is the production of honey for types *H. itama*.

2016, began trials with resin-dependent species.

Meliponicultursts (Stingless Beekeepers) in Brunei Darussalam

The total number of players in each village ads up to 262 stingless beekeepers in Brunei Darussalam, Whatsapp Lebah Kelulut Brunei, 156 members as of 1st May 2016.

Figure 100 a) b) f) Collection of box hive designs d) dark honey c) e) H. itama hive giving dark honey i) Collection of different colours of meliponine honey.

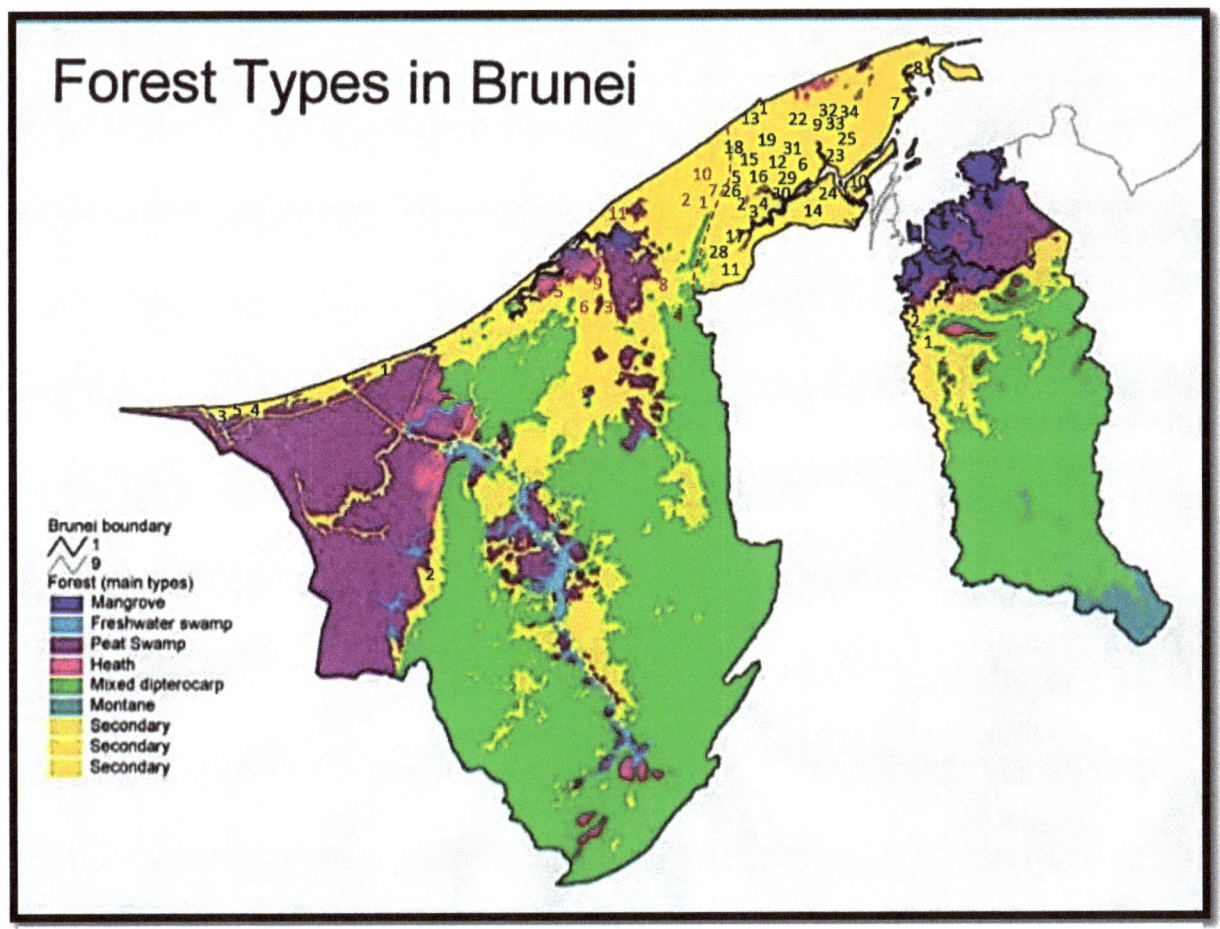

Map 1 Map of Stingless Beekeepers by Village and forest types in Brunei

Daerah Brunei Muara:
1. Kg Tungku - 80 keepers
2. Kg jangsak - 4
3. Kg batumpu masin – 2
4. Kg bengkurong – 5
5. Kg Sinarubai – 1
6. Kg Burung Lepas – 6
7. Kg pasai -5
8. Kg beribi -4
9. Kg Pangkalan sibabau -1
10. Kg perpindahan serasa -1
11. Kg Kapok -3
12. Kg. Menunggol -1
13. Limau Manis – 10
14. Kg Mata-mata : 9
15. Kg Jerudong : 21
16. Kg kasat -1
17. Kg tanjung nangka – 5
18. Kg panchor murai -8
19. Kg batong – 1
20. Kg katimahar – 3
21. Kg Selayun 'B' – 6
22. Kg rimba-15
23. Kg subuk-3
24. SKPRRJ Bukit Sebanging – 1
25. Kg lumapas-3
26. Kulapis – 2
27. Berakas kem – 1
28. Kg Batu Ampa – 1
29. Kg bebatik kilanas-1
30. Kg kiarong-1
31. Kg jaya setia perpidahan berakas-3
32. Kg lambak-2
33. Kg panca delima-1
34. Kg madang-1
35. Kg mulaut ban3-2
36. Kg kilanas-1
37. Kg masin-1

Daerah Tutong:
1. Kg Sg Kelugos – 2
2. Kg sinaut – 5
3. Kg bukit – 1
4. Kg lamunin – 13
5. Kg Beruang – 7
6. Kg Penapar -1
7. Kg Kupang – 1
8. Kg Kiudang – 1
9. Kg tanjung maya – 1
10. Kg Keriam – 2

Daerah Belait :
1. Kg lumut -1
2. Kg rempayoh – 1
3. Kg mumong – 1
4. Perumahan 2000 panaga -1
5. Kg pandan-1

Daerah Temburong:
1. Kg rataie-6
2. Kg ujung jalan-1

Table 1 Table of Villages with no. of Stingless beekeepers in Brunei

Wild Bees of Brunei Darussalam David Roubik 1996v

Table 1. Highly social bees and nests encountered near the Belalong field station.

Species	Nest (trees, No.)	Nest height (m)
Apis koschevnikovi 'ranyuan'	-	-
A. dorsata 'manyi' or 'lanyeh'	*Koompasia*, 1	40
A. cerana 'ranyuan tikung'	-	-
A. andreniformis 'nuang'	-	-
Pariotrigona pendleburyi	*Shorea, Dryobalanops*, 8	0.04-0.4
Trigona (Lepidotrigona) nitidiventris	-	-
T. (L.) terminata	1	4
T. (L.) ventralis	1	5.5
T. (Heterotrigona) melina	2	ground level
T. (H.) haematoptera	1	8
T. (H.) hobbyi	-	-
T. (H.) collina	11	ground level
T. (H.) thoracica	-	-
T. (H.) apicalis	3	1.5-3
T. (H.) drescheri	1	ground level
T. (H.) melanocephala	3	0.1-0.9
T. (H.) reepeni	-	-
T. (H.) itama	-	-
T. (H.) erythrogastra	-	-
T. (H.) canifrons	-	-
T. (H.) fuscobalteata	-	-
T. (H.) fuscibasis	-	-
T. (H.) rufibasalis	-	-
T. (H.) geissleri	2	0.2-0.5
T. (H.) laeviceps	-	-
T. (Homotrigona) fimbriata	-	-

Table 2 List of Species of Wild Bees found during the Tropical Rainforest Research 1993 in Brunei

Reference:

Roubik, D. W. (1993). Wild bees of Brunei Darussalam. In D. B. Edwards, *Tropical Rainforest Research — Current Issues Volume 74 of the Monographiae Biologicae* (pp. 59-66). Bandar Seri Begawan: Springer.

PART 2
~ INDONESIAN ARCHITECTURE ~

Introduction to Part 2

Part 2 explores the ethnic cultures and architecture in Indonesia. It starts with Sumatra being the largest land mass with umpteen diverse cultures and spiritual faiths. The North-western regions, i.e., the Bataks, Acehnese, and the Minangkabau of West Sumatra. From there, this part also covers Central and South Sumatra.

In Lampung, S. Sumatra, Heri Damora, the owner of Al Qorni Bee farm, is an avid beekeeper with a penchant for miniature model making. Many of his works are featured herein for readers who have a keen interest in miniature model architecture. Traditional model houses are great when preparing a Meliponarium (a Terrarium that includes feral colonies in rotted hollow branches). In these terrariums, one places small model houses to attract the bees to multiply, and they serve as bait boxes.

A look at the Kampar Majo tribe with their Malay Lontiak House. The Lontiak house is the most aesthetically pleasing among all Malay-styled traditional houses. Here, I scrutinize the ethnic houses of the Sasak tribe and the grain store called Lumbung and how the Dutch colonial barn-type structures have influenced the structure. Also, I take a glance at traditional Indonesian Wallacea like Lombok.

Javanese Architecture in Meliponiculture has much influence from the Malay Nusantara traditional architecture as in the Sundanese Architecture of West Java. The variations in the Javanese houses in different regions in west and central Java show significant development in the vernacular designs

Figure 101 Some assorted vernacular architecture in the Indonesian Archipelago, Compilation of ethnic houses from an Indonesian Tourism website.

between the Betawi cultures and the Baduy (a Sunda sub-ethnic group). I compare these ethnic influences in the Javanese tribes of the Sunda, Betawi and Baduy peoples.

Chapter 5

The Bataks of Sumatra

The culture with the ancestral veneration emphasis is the Toba Batak of Samosir Island in Lake Toba, Sumatra (an island in a lake on an Island). The Batak are often considered isolated people, thanks to their location inland.

Figure 102A Toba Batak house and a model Batak roof of a stingless bee box attached to a log hive.

They practised a syncretic religion of Shaivism, Buddhism and local culture for thousands of years.

This roof is a typical Batak house with a saddle roof (concave ridge) similar to the Minangkabau, but the pitch and the front tip are usually higher than the back tip. The simulated air movement shows the ultimate in heat dissipation. Below are some variations of the saddle roof used for bee housing.

Figure 103 The upswept ridge of a centra Sumatra house model and the Acehnese house model of a bee box hive. On the right is a structure of a saddle roof model usually found in Sumatra.

Batak Karo of North Sumatra

Batak Karo is the assimilation of the Batak people with the Karo people. Batak Karo traditional house is one of North Sumatra's traditional houses, also often called the Siwaluh Jabu traditional house. Decorticated from the name, Siwaluh Jabu is a house that eight families inhabit. And every family has their respective roles in the house.

The Karo, or Karonese, are a people of the Tanah Karo (Karo lands) and part one of the Batak people sub-ethnic

Figure 104 Batak Karo House, N. Sumatra

group from North Sumatra, Indonesia. The Karo lands consist of *Karo* Regency, plus neighbouring areas in East Aceh Regency, *Langkat* Regency, *Dairi* Regency, *Simalungun* Regency and *Deli Serdang* Regency. In addition, the cities of Binjai and Medan, bordered by Deli Serdang Regency, contain significant Karo populations, particularly in the Padang Bulan area of Medan. The Sibolangit, Deli Serdang Regency, in the foothills of the road from Medan to Berastagi, is also a significant Karo town. Karoland contains two major volcanoes: Mount Sinabung, which erupted after 400 years of dormancy on 27 August 2010 and Mount Sibayak. Karoland consists of the cooler highlands and the upper and lower lowlands. Source[20]

Figure 105 Left: Traditional longhouses at a Karo village near Lake Toba, circa 1870; Middle: A Karo people church affiliated with Karo Batak Protestant Church (GBKP). Kabanjahe, Karo Regency, North Sumatra; Right: Musical instruments and other items identified as Karo Batak, photograph (circa 1870) by Kristen Feilberg. <:

The religion of the Karo people is mostly Christian, a religion brought to Sumatra in the 19th Century by Dutch missionaries. However, a populace away from the Karo Highlands has converted to Islam, with the influence of Muslim Malays and Javanese immigrants, thus making the traditional habits of pig farming and pork dishes uncommon. Some Muslims and Christians, however, still retain their traditional animist beliefs in ghosts and spirits (*perbegu*) despite contradictions to their other beliefs. Their traditional longhouses, with few structures left nowadays, are used for ceremonies and harvest festivals called '*Kerja tahun*' and occasionally a *perbegu* 'calling' ceremony. The *perbegu* ceremonies start with the traditional folk music ensemble chanting the following snippet[21] by the leading 'shaman'.

Gayo Nusantara - Ethnic Gayo Tribes in Dataran Tinggi, Tanoh Gayo – Indonesia.

---

[20] https://en.wikipedia.org/wiki/Karo_people_(Indonesia)
[21] Enko ise. (Kau siapa) Who are you? Arkai kam ije (ngapis kau kesini) what are you doing here?
Kai dahin mu ijenda (apa kerja mu) what do you do? Ras ise kam ije (sama siapa kau kesini) who are you here with? ... goes on with mich fanfare, and finally with Majuajua Hora. Greetings or Welcome in the Karo language.

Batak Gayo

The Gayo tribe is a tribe that inhabits the Gayo highlands in Aceh. Most of the Gayo tribe are found in Central Aceh, Bener Meriah, and Gayo Lues districts and three sub-districts in East Aceh, namely Serbe Jadi, Peunaron and Simpang Jernih districts. In addition, the Gayo tribe also inhabits several villages in the districts of Aceh Tamiang and Aceh Tenggara. The majority of the Gayo tribe is Muslim. The Gayo tribe uses a colloquial language called the Gayo language. The Gayo tribe is different from the Acehnese. Physically, the language and culture of the Gayo tribe are more like the Alas, Singkil, Kluet, Karo Batak and Pakpak Batak tribes.

Gayo people[22] live in small communities called kampongs. A gecik heads each village. A collection of several villages is called a *kemukiman*, led by a mukim.

Gayo Lues tribe in Gayo Lues district

The Gayo Lues tribe is a Gayo sub-tribe that lives in the Gayo Lues district and several sub-districts in Southeast Aceh. The settlement of the Gayo Lues tribe, which is in the Gayo Lues district, is in the Bukit Barisan Mountain range, most of which is an isolated area of the Gunung Leuser National Park in the province of Aceh. Also, a small part is in South Aceh, Aceh province.

Figure 106 Traditional House of The Gayo Lues tribe

The culture and customs of the Gayo Lues sub-ethnic have almost no differences from other Gayo sub-tribes, such as Gayo Serbejadi (Lukup), Gayo Kalul, Gayo Lut and Gayo Deret. It's just that they are distinguished from the dialect used. They have a different dialect from other Gayo sub-languages.

Even though they have a different dialect from other Gayo groups, they are not a different tribe from other Gayo tribes. They are still Gayo tribes. Maybe because their territory is different and separate from other Gayo ethnicities, and the dialect they speak is slightly different, they are called Gayo Lues. The Gayo Lues people, the majority of whom are Muslim, were brought by the Acehnese and Minangkabau people, whose descendants also live in this area. They are devout Muslims, so some of

---

[22] https://en.wikipedia.org/wiki/Gayo_people

their cultures contain many Islamic elements. Gayo Lues people generally live in agriculture, such as growing vegetables, red chillies, citronella, cocoa, tobacco and arabica coffee.

Gayo Kalul in Aceh Tamiang

Gayo Kalul[23] people, sometimes called Gayo Kaloy people, are a Gayo ethnic that lives in Kabupaten Aceh Tamiang. The culture and customs of the Gayo Kalul people are different from other Gayo people because the location is separated in the western part of Aceh Tamiang. It is influenced by the culture of the Melayu (Malay) Tamiang people because of the intense interaction between these Gayo people with them. Generally, Gayo culture is similar to Batak Alas, Batak Singkil, Batak Gayo and Batak Karo. Nowadays, their culture is mostly influenced by Islamic culture.

The culture and customs of the Gayo Kalul sub-ethnic are almost no differences from other Gayo sub-tribes, such as Gayo Serbejadi (Lukup), Gayo Deret, Gayo Lut and Gayo Lues. It's just that they are distinguished from the dialect used. They have a different dialect from other Gayo sub-languages. Some words from the Gayo Kalul language have some differences but can still be understood by other Gayo tribes, for example, in referring to people, whereas in the Gayo language, the word "people" is generally "*jema*", while in Gayo Kalul it is "*urang*". Although there are some differences in the vocabulary of the Gayo Kalul language, they can still communicate with other Gayo tribes.

The Gayo Kalul people live as farmers in the fields and gardens around their residential areas. Currently, many of them have worked in the private and government sectors. Few of them began to migrate to other areas, such as Banda Aceh, Medan and the island of Java.

Batak Bebesen in Aceh Tengah

The Batak '27' tribe, or the Bebesen Batak[24] tribe, is a group of people who are in the customs of the Gayo Batak tribe, which formerly came from the North Batak lands and migrated to the Gayo Land area. This Batak '27' tribe lives in the Bebesen area, which is still part of the Gayo tribal area.

In the past, many Batak people from the North came to Gayo Land in various ways, now living in the west of Laut Tawar Lake, their descendants cannot be distinguished. However, one memory still lingers in the minds of the Gayo people, namely what happened to Reje Cik Bebesan's men and Ketol. The next is the descendant of one of the famous and prominent Reje in Gayo Land who inhabits the eastern part of the Jemer watershed, namely *Reje Linge*. The Gayo people in question are the Gayo

---

[23] https://ms.wikipedia.org/wiki/Gayo
[24] https://ms.wikipedia.org/wiki/Bahasa_Gayo

Bebesen people who live in the western part of Lake Laut Tawar. When fighting with their neighbouring villages, the Batak Bebesan or Batak 27 often ridiculed them.

In folklore, 26 men from North Batak defeated Reje Bukit's troops in a battle. Reje Bukit himself, in a state of panic, fled and got lost in a swamp (swamp) near the village of Kebayakan, so the place is called Paya Reje to this day. After that, an agreement was made which stated that Reje Bukit and his men were appointed to occupy the village of Kebayakan, and the 26 North Batak people, all of whom had converted to Islam, occupied the area that has now become the mother village of Raja Cik, namely the Bebesen village. In this place, they thrive, and their descendants are called the Batak Bebesen tribe or the Batak 27 tribe.

Batak Gayo Serbejadi in Aceh Timur

The Gayo Serbejadi tribe (Lokop, Lukup) is a Gayo sub-tribe that lives in the East Aceh district of Aceh province. It is said that the Gayo Serbejadi tribe is a community group originating from the Gayo Lues, Deret and Lut people, who migrated to this area. Living for centuries together to form a community slightly different from other Gayo ethnicities, they are referred to as the Gayo Serbejadi or Gayo Lukup ethnicity. They have a slightly different culture from other Gayo groups. But even so, they still recognize themselves as Urang Gayo, with the embellishment of Serbejadi.

Figure 107 Traditional House of Gayo Serbejadi

Most of the Gayo Serbejadi people[25] are Muslims. They are devout Muslims. So that all the culture and customs that they practice contain Islamic elements. Even though they are devout in their religion, they accept all people from different religions to live side by side in their territory. Their residential areas are currently being entered by migrants from various regions in Sumatra, even from outside Sumatra.

Gayo Serbejadi people generally live as farmers, farming and growing vegetables, red chillies, cocoa, tobacco, and Arabica coffee scattered in various residential areas of the Gayo Serbejadi tribe.

---

[25] https://commons.wikimedia.org/wiki/Category:Gayo_people

Batak Gayo Deret in Aceh Tengah

The Gayo Deret[26] tribe, also known as Gayo Linge, is a Gayo sub-tribe that lives in the Linge area and its surroundings (still a part of the Aceh Tengah district in Aceh province. The culture and customs of the Gayo Deret sub-tribe are almost no differences from the Gayo Deret tribe. However, other Gayo sub-tribes, such as Gayo Serbejadi (Lukup), Gayo Kalul, Gayo Lut and Gayo Lues, differ from the dialect used. They have a different dialect from other Gayo sub-languages.

In the Gayo Deret area, a large kingdom once stood around the 10th century called the Linge Kingdom (Kingdom of Linga). The Gayo people founded the Linga Kingdom in the past, whose first king was Genali. It is said that the Gayo people converted to Islam before the Acehnese, who became the majority in the province of Aceh.

There are only differences in terms of the mention of some cultural terms and their customs. According to them, the Gayo Deret people and other Gayo clans are the same only because the different regions separate them. They are called Gayo Deret. But some parents from the Gayo Deret community said that Deret is the name of someone their King previously assigned to care for all kinds of animals in their customary territory.

Figure 108 Traditional House of Batak Gayo Deret

The Gayo Deret community initially lived in agriculture, cultivating and growing various vegetables and fruits. Currently, they are trying more perennials such as coffee, cocoa and others.

Batak Gayo Lut in Bener Meriah

Aceh Tengah district, Aceh province. The Gayo Luttribe is a Gayo sub-tribe around the Laut Tawar Lake. The Gayo Lut tribe is known as Gayo Lut because their area of residence is around Lake Laut Tawar, which in the Gayo language is called Lake Lut Tawar. Besides being called Gayo Lut, they are sometimes known as Gayo Laut.

---

[26] https://ms.wikipedia.org/wiki/Gayo

The Gayo Lut community, the majority of whom are Muslims. The settlements of the Gayo Lut tribe formerly consisted of houses on stilts that could reach 20 to 30 meters in length and 10 meters in width. At the bottom is a place to store livestock, such as cows and goats. Islam has long developed in this area. It is said that according to their story, Islam entered the Gayo community first, then the Acehnese tribe.

In general, the life of the Gayo Lut community is by profession as a farmer, such as planting rice in the fields, farming, and growing various vegetables. They also grow perennial crops such as arabica coffee, which is growing and popular, such as Gayo coffee. In addition, some of them live as fishermen in the Laut Tawar lake. Few Gayo Lut people have succeeded overseas, becoming entrepreneurs or government officials.

Art is a never-sluggish cultural element among the Gayo people, which rarely stagnates and even tends to develop. The famous Gayo art forms include the saman dance and the art of speech called didong. In addition to entertainment and recreation, these art forms have the function of ritual, education, and lighting, as well as maintaining balance and the social structure of society. In addition, there are also art forms such as *Bines* dance, *Guel* dance, *Munalu* dance, *Sebuku* (pepongoten), guru didong, and melengkan (the art of speech based on custom), which are also not forgotten from time to time.

Batak Kluet[27] people are related to Batak Alas, Batak Gayo, Batak Singkil and Batak Karo because they have similar languages, customs, and cultures. Even though they live in Aceh Province, their cultures are closer to Batak than Aceh. Besides that, they have similar clans, such as Monte in Batak Kluet, which have the same ancestor as Munthe in Batak Alas, Munte in Batak Gayo, and Ginting Munthe in Batak Karo.

Figure 109 Traditional House of The Batak Kluet tribe

Bataks in Aceh Province [28]

These are some of the names of North Sumatran traditional houses that are rarely known.

---

[27] https://ms.wikipedia.org/wiki/Bahasa_Batak_Alas-Kluet
[28] Reference:
1. http://id.wikipedia.org/wiki/Kabupaten_Aceh_Selatan
2. http://id.wikipedia.org/wiki/Suku_Kluet
3. http://protomalayans.blogspot.com/2011/07/kluet.html
4. http://yasirmaster.blogspot.com/2011/11/kluet-tenggelam-dalam-sejarah.html
5. http://dmilano.wordpress.com/2011/03/27/suku-kluwat-dan-sejarahnya/

Figure 110 Traditional House of Angkola Batak tribe

1. Rumah Adat *Angkola*

Many equate the *Angkola* Batak tribe with the Mandailing Batak tribe, even though the two are not the same and have different traditional houses. The traditional *Angkola* house, Bagas Godang, is made of plank walls, floors, and palm fibre roofs and is dominated by black. The uniqueness of the *Angkola* traditional house is a characteristic that distinguishes it from other traditional houses.

2. Rumah Adat *Simalungun*

The main characteristic of the Simalungun traditional house that is different from other North Sumatran traditional houses lies in the shape of the roof. The Simalungun traditional house is named *Bolon* House, which is like the name of another traditional house, namely the *Bolon* house. Another characteristic of the Simalungun traditional house lies at the foot of the building made of logs that cross between the corners.

Figure 111 Simalungun traditional house

Figure 112 The Mandailing traditional house

3. Rumah Adat Mandailing

The Mandailing traditional house is inhabited by the Mandailing tribe living on the Riau Province's border. In the local language, the Mandailing traditional house is called *Bagas Godang,* where Bagas means house and Godang means many. Structurally, the Mandailing traditional house has a shape that is quite different from other North Sumatran traditional houses and is the main feature.

4. Rumah Adat Pakpak

Figure 114 The Pakpak/Dairi traditional house

In general, Pakpak traditional houses have characteristics and shapes that are not much different from other traditional houses. The Pakpak/Dairi traditional house has a distinctive shape; the building is made of wood, and the roof is made of palm fibre. The traditional house, also known as *Jerro*, represents the typical Pakpak culture in every part of the building.

5. Rumah Bolon

Rumah Bolon[29] is an iconic traditional house of North Sumatra whose uniqueness has been recognized nationally. The Bolon house[30] is also a symbol and identity for the Batak people, rich in history and philosophy.

Figure 113 The Bolon house is a symbolic identity of the Batak people

The shape of this house is rectangular, with all parts of the house made of materials derived from nature. This stage-shaped house is usually used for family gatherings and traditional events. Generally, Bolon traditional houses are inhabited by 4-6 families living together and some livestock are kept under the house.

Figure 115 A model of a traditional jambur

A ***jambur***[31] is a structure used as a multipurpose hall by the Karo people of North Sumatra, Indonesia. The traditional jambur is a large pavilion-like structure, raised above the ground, wall-less, and placed under a large Karo traditional house roof style. Karo ritual ceremonies, e.g., wedding feasts, funerals, or general feasts, are held within the jambur. Jambur can still be found in big cities of North Sumatra, e.g., Medan, Kabanjahe and Berastagi, as well as small villages in the Karo land.

---

[29] References:
https://rimbakita.com/rumah-adat-batak/
https://www.99.co/blog/indonesia/rumah-adat-sumatera-utara/
https://elib.unikom.ac.id/files/disk1/679/jbptunikompp-gdl-iasantipur-33904-2-unikom_i-i.pdf
[30] https://en.wikipedia.org/wiki/Bolon_house
[31] https://en.wikipedia.org/wiki/Jambur

Distribution of Meliponiculture with Vernacular architecture

Figure 116 Vernacular architecture in Sumatra, Indonesia.

Many have asked, while the range of models is diverse and pretty, are they really practical to be converted to house bees? To answer this, attached are two videos that speak for themselves.

1. https://www.youtube.com/watch?v=QW1oAF7G9Co
2. https://www.youtube.com/watch?v=ndkONWvem3o

The first video shows a Tetragonula sp. in a Batavian house model.

It takes a load of skill and experience for even the miniature model hobbyist to muster. Done well; it keeps the beekeeper happy and the bees happy. Contentment in conservation is sowing the seed that grows into a tree that provides shade for all but not necessarily the planter himself.

An interesting link is https://www.dezeen.com/2022/01/24/bee-bricks-planning-requirement-brighton/, which should be shared in the context of Vernacular Architecture, regardless of where one is. A matter of innovation and ingenuity

Chapter 6

Aceh and the North Sumatra Architecture

Figure 117 Rumoh Aceh image source: Wikimedia

Rumoh Aceh is made entirely of wood, without nails. Traditionally, the floors are made of feather palm planks, the walls of thin woven bamboo, and the roof of thatched sago palm leaves. The entire construction is erected over a pile construction which stands on stones. The ground under the house is compacted and made a bit higher than the area around the house. The soil is prevented from seeping away by edgings around this compacted soil. In the colonial era, the edgings were made of bottles planted in the ground. Source[32]

In the Aceh matriarchal lineage, when a daughter reaches the age of seven, her father will start collecting building materials to construct the house where his daughter will live with her future husband. When her parents die, the daughter will acquire the rice fields and her parents' house. According to Acehnese custom, the girl must live with her husband in her mother's house until the first child is born. Afterwards, she can move to her house, within the parent's compound. When her parents die, the daughter will acquire the rice fields and the parent's house.

Unlike the traditional houses of the Balinese, the Rumoh Aceh is on the verge of extinction due to the lack of the good timber required for their construction and their impracticality for the modern lifestyle. The most reasonable reason is perhaps an interest in change: people living in traditional houses are associated with poverty, compared to the seemingly higher status of modern people living in modern houses.

Acehnese traditional house in Piyeung Datu village[33], Montasik district, Aceh Besar regency.

---

[32] https://en.wikipedia.org/wiki/Rumoh_Aceh#:~:text=Rumoh%20Aceh%20is%20a%20pile,palm%20leaves%20or%20corrugated%20metal.

[33] https://upload.wikimedia.org/wikipedia/commons/1/1d/Rumoh_Ac%C3%A8h_di_Piyeung_Datu.jpg

Several Acehnese architects have attempted to modernise the traditional architecture of Aceh. One attempt is to replicate the roof canopy of the Cakra Donya bell (an ancient relic in Aceh) in the State Museum of Aceh auditorium. Wim Sutrisno, a local architect, completed this reinterpretation of the Cakra Donya roof. The Cakra Donya roof style has become popular in Aceh, decorating hotel gates or bus shelter roofs. Many buildings have reinterpreted the gable roof,

Figure 119 Acehnese traditional house in Piyeung Datu village, Montasik district, Aceh Besar regency. Source: https://upload.wikimedia.org/wikipedia/commons/1/1d/Rumoh_Ac%C3%A8h_di_Piyeung_Datu.jpg

Figure 118 In the background, Aceh Museum featuring the traditional Rumoh Aceh is a modern interpretation of the Cakra Donya roof. Source: https://en.wikipedia.org/wiki/Rumoh_Aceh#/media/File:Museum_Aceh.JPG

In the background, the Aceh Museum featuring the traditional Rumoh Aceh is a modern interpretation of the Cakra Donya roof.

https://en.wikipedia.org/wiki/Rumoh_Aceh#/media/File:Museum_Aceh.JPG

The Aceh Escapade
by Abu Hassan Jalil

On the left photo is an Aceh Bee Box Hive designed by Omjol of Peureulak, East Aceh.

Figure 120 On the left photo is an Aceh Bee Box Hive designed by Omjol of Peureulak, East Aceh. The photos on the right are those of the Rumah Adat, a communal house at the museum in Banda Aceh, North Sumatra.

The photos above on the right are those of the Rumah Adat, a communal house at the museum in Banda Aceh, North Sumatra. A recent webinar at The Indonesian Honey Festival in Banda Aceh presented the two proposals below.

Figure 121 These are adaptations of modern architecture; one was at their museum, and the other is a Belltower.

These are adaptations of modern architecture; one was at their museum, and the other is a Belltower.

Mandailing Natal Regency

Bagas Godang Panyabungan Tonga[34] is a traditional house of the Mandailing, one of the indigenous tribes of North Sumatra, and is often grouped as one of the ethnic groups of the Batak Tribe. It is said that once upon a time, *Bagas Godang* was the residence of the Panyabungan kings, whose surname was Nasution.

Figure 122 Bagas Godang in Panyabungan, Mandailing Natal

---

[34] Source: https://en.wikipedia.org/wiki/North_Sumatra#/media/
File:Bagas_Godang_Panyabungan_Tonga_(Batak_Mandailing_House)_(04).jpg

Minangkabau of West Sumatra

by Abu Hassan Jalil

A significant SB species associated with the Minangkabau people is *Tetragonula* **minangkabau** (Sakagami & Inoue, 1985) – The epithet relates to an ethnic people or the highlands of Sumatra Indonesia. Typ. Loc.: Indonesia "Lubuk Mintrun nr. Padang, Sumatera Barat, Indonesia,] (Rasmussen, 2008)

Figure 123 AHJ upon arrival at the Minangkabau airport in Padang, West Sumatra

When in Padang some five years ago, the first picture I was compelled to take was the architecture of the Airport with the Minangkabau signage.

Though my mission was to observe the *Tetragonula minangkabau* stingless bee, I could not resist being engrossed in the complexity of the roofing structure. Many have seen this architecture but have not been impressed by the complicated construction of its concave ridges and interwoven rafters.

Prof Dr. Siti Salmah told me about her understudying the late Prof Sakagami (sensei). They were observing a stingless bee colony developing. The bees flew between the mother colony and the new nest (about 50 meters away). They gathered building materials and provisions from the mother colony to equip the new nest. Prof Siti said she mentioned to Prof Sakagami that she did the same with her parent's house.

When she got married and, in her culture, as newlyweds, she was allowed to travel frequently to her parent's house to gather light furniture, curtains, kitchen utensils and other necessities to furnish her new house. It is like the bees' behaviour that Sakagami named the bee after the people of that culture.

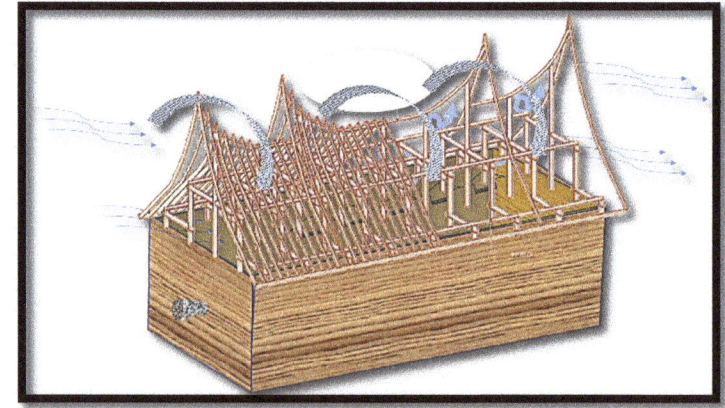

Figure 124 Simulated air movement on a skeletal Minangkabau 'Rumah Gadang' roof design on a box hive model.

The people are champions at breeding buffalos and always win regional buffalo competitions. The word win is 'menang', and buffalo is 'kerbau' in Malay. Then the term 'menang kerbau', colloquially pronounced

Minangkabau, became the name of the people, culture, and even the Airport. Henceforth the stingless bee was named *T. minangkabau* by Sakagami.

The architecture undoubtedly bears the horn of the buffalo. The ends may also be tipped with a spire or finial that may serve as lightning arresters. In the old days, the spire may have been called '*tunjuk langit*', meaning point to the sky.

Anyway, back to the architecture and a look at the air movement the roof structure induces when adopted onto a stingless bee box hive:

Besides the *Rumah Gadang*[35] Minangkabau, Architecture variations in the West Sumatra houses exist in the *rangkiang*[36] rice granary.

Figure 125 Left: A model bee box hive roofing; Middle: Two rangkiang c. 1895 rice granaries; Right: A model of a rangkiang

The bottom left is a Bee box hive in the replica of the *rangkiang*.

This house was built with long pillars that rose and were resistant to shocks. It has 4 main pillars of Juha trees (*Senna sumatrana*). Before being used as a pole for this house, the Juha tree was soaked in a pond for years to produce a strong and sturdy pole. This pole has a diameter of 40cm to 60cm. A rangkiang is a structure built over a raised pile foundation. It has a distinguished roof shape known as a *gonjong* ("spired") roof, like a Minangkabau traditional house, the *Rumah gadang*. The *gonjong* roof symbolically identifies it with buffalo horns. Like the *Rumah gadang*, a rangkiang is traditionally a thatched roof made of palm fibre (ijuk) and is also similarly decorated. The only opening to a rangkiang is a small rectangular hatch high up in the roof, where the harvested rice is placed. A ladder is required to reach this hatch.

---

[35] https://en.wikipedia.org/wiki/Rumah_Gadang
[36] https://en.wikipedia.org/wiki/Rangkiang

A *balairung*[37] is a village hall of the Minangkabau people of West Sumatra, Indonesia. It has a similar architectural form to the *Rumah gadang*, the domestic architecture of the Minangkabau people. Whereas a Rumah gadang is a proper building, the balairung is a pavilion-like structure used solely for holding a consensus decision-making process in the Minang society.

Figure 126 Left: A wall-less balairung in Batipuh.' Right: A balairung in Matur

The **Mentawai**[38] Islands Regency are a chain of about seventy islands and islets approximately 150 kilometres (93 miles) off the western coast of Sumatra in Indonesia. *Uma*[39] is a traditional vernacular house found on the western part of the island of Siberut in Indonesia. The island is part of the Mentawai islands off the west coast of Sumatra.

Figure 127 An Uma, the traditional communal house of the Mentawai

The Acehnese style influences the structures built on a much larger scale. They were formerly used as uma longhouses by the Sakuddei tribe. Uma longhouses are rectangular, with a veranda at each end. Built on piles, they traditionally have no windows. They can be 300 m2 in area.

List of records of Stingless bees found in Sumatra (Rasmussen 2008)

Sundatrigona lieftincki (Sakagami & Inoue, 1987):
Trigona (Trigonella) lieftincki Sakagami & Inoue 1987: 610, 611, 613-615, 617-624: Holotype (RMNH, not located, nor in SEHU); Paratypes (RMNH, three workers, two males) (morphology, taxonomy); **Type locality**: INDONESIA "N.E. **Sumatra**, Tongkoh, probably Mt. Talamau (Pasaman), VI 1941, v.d. M. Mohr leg." (5 workers, four males, holotype

Tetragonula minangkabau (Sakagami & Inoue, 1985)
Trigona (Tetragonula) minangkabau Sakagami & Inoue 1985: 175, 176, 177, 178, 179, 180, 181, 184-186*, 187-188: Holotype (MZB, Hymn.0198, worker); paratypes (MZB (Hymn.0195-0197)) (key to species, nest, taxonomy); **Type locality**: INDONESIA "Lubuk Mintrun nr. Padang, **Sumatera Barat,** Indonesia, ix 1981 in mass flight, S.F. Sakagami" (1 worker, holotype); "Lubuk Mintrun" (many workers, 25 males, paratypes); worker, paratypes); "Koto Atas, Solok, Sumatera Barat" (6 workers);

---

[37] https://en.wikipedia.org/wiki/Balairung
[38] https://en.wikipedia.org/wiki/Mentawai_Islands_Regency
[39] https://en.wikipedia.org/wiki/Uma_longhouse

Chapter 7

Vernacular architecture in Central and South Sumatra

Malay Lontiak House of Kampar Majo Tribe

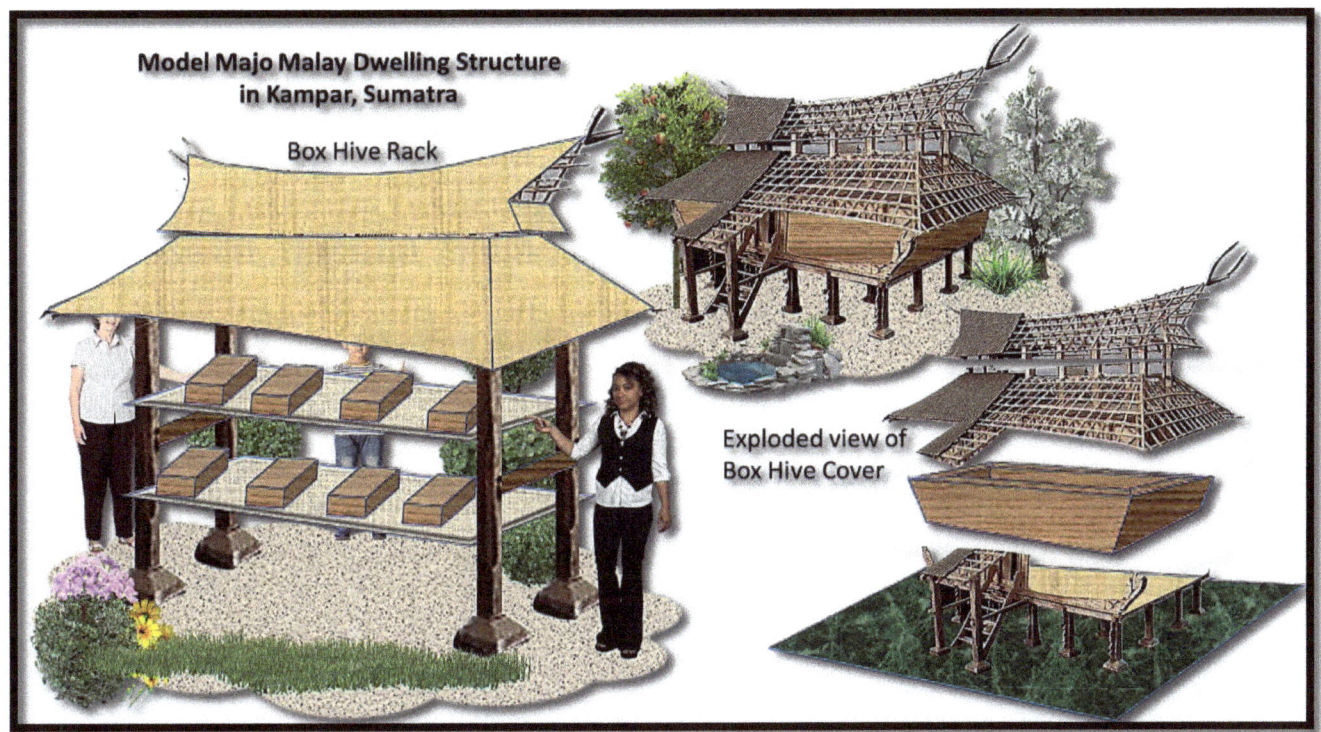

Figure 128 Replica of Malay Lontiak House of Kampar Majo Tribe and applied on the roof of a bee box hives rack

Malay architecture has a lot of typologies in roof forms, such as Limas, Lontiak, Begonjong, Layar and Sayap, Peranakan (a mixture of ethnic Chinese, see *Rumah Agam* p.40), and several other typical Malay houses. Majo Melayu House is a stage building with the characteristics of the Lontiak roof. In general, this house is divided into 2 (two) building parts: the first line is the main house, and the second building mass is the kitchen. There is a connection between the main house and the kitchen. The ornaments first seen in this house are *Selembayung* or *Tanduk Buang*; there were also ornaments such as unsheathed spears called *Tombak-tombak* and kite wings *Sayap Layang-layang* on the four corners of the roof. Various types of carvings are found in every segment of this building.

1. *Rumah Lontiok* (meaning Lontiak House, see also *Rumah Lontik* p.37) is a traditional Kampar, Riau Province house. *Lontiok* (*lentik* in Bahasa Indonesia or *lontik* in Minang-Malay) means arched or curved. The Lontiok term corresponds to the form of the roof ridge, which is arched upward as a symbol to honour Allah. Other terms used for *Rumah Lontiok* are *Rumah Pencalang* or *Rumah Lancang* (traditional sailing wooden ship). It is based on the similarities between the base of the front and back of the wooden wall with Lancang (the wooden sailing ship), the two similarities between the form of the outer leaning of the wooden wall of Lontiok

House with Lancang, which is leaning outward, and the similarities between Lontiok House with traditional wooden sailing ship house (*magon*) made by the community living onto the river and sea[40].

2. *Rumah Bubung is a traditional house from the Riau archipelago in Indonesia (See also Rumah Bubung p.37)*. The Belah Bubung house is also known as the *rabung* house or the Malay bubung house. It is said that the name of this house was given by foreigners who came to Indonesia, such as China and the Netherlands.

3. *Rumah Lamo* - The first people who lived in the Inner group, Inner residential people called *Kajang Lako* or Old House (See also *Rumah Kejang Lako* p.37). *Lamo* ridge forms such as a boat house with the upper end of the ridge upward. Lamo house typology shaped hall, rectangular with a length of 12 m and width of 9 m. the rectangular form is intended to facilitate the preparation of a room adapted to its function and is also influenced by Islamic law.

4. *Rumah Limas* is a traditional house in Palembang, South Sumatra Province (See also *Rumah Limas* Johor p.33). Called the pyramid house because of its shape, which resembles a pyramid. The pyramid house is a multi-storey building, and each level has meaning. The community of Bengkalis calls the levels in this house. The pyramid house is very large and is often used as a venue for weddings and traditional events. They range from 400 to 1000 sq. meters.

5. The Bengkulu traditional house has another name, Bubungan Lima. It is so named because of its tiered or soaring roof. The uniqueness of the Bengkulu traditional house is the shape of the roof, which is stacked and made of palm leaves. In addition to the unique shape of the house, this house also has an earthquake-resistant design because it is supported by 15 poles that are 1.8 meters high.

6. *Rumah Adat Nuwo sesat* is one of the traditional houses in Lampung Province. *Nuwo Sesat* is a traditional meeting place for purwatin (balancing) when holding traditional pepung (*Musyawarah*). Therefore, this traditional house is also called *Balai Agung*. The parts of this traditional house are the Pavilion, a foyer used for small gatherings. The *Pusiban* is an inner room used as a place for official deliberations. The *Tetabuhan* room is a room for storing traditional musical instruments, the *Gajah Merem* room is used for resting places for balancers,

---

[40] For further reading: "Lamgkau Bentang" Jurnal Arsitektur, Vol.6, No.1, 2019. Wood Carvings on a Majo Malay House in Kampar, Sumatra. Personal sketches by Gun Faisal, 2019.

and the *ijan* deck is an entrance equipped with a roof. The roof of this traditional house is called *Rurung Agung*.

Lampung traditional houses[41]

Figure 129 Lampung traditional house named Nuwou Sesat

The Lampung traditional house named *Nuwou Sesat* comes from 2 words: "*Nuwou*", which means house, and *Sesat* means custom. The main function of the *Nuwou Sesat* traditional house is to serve as a hall or gathering place for all residents. However, besides being a gathering place for residents, Nuwou Sesat is a house on stilts because many rivers drain the area. How to make it follow the river's flow with a tight pattern.

In the form of houses on stilts, traditional houses are also useful for avoiding wild animals. The building is also made strong and earthquake-resistant because Lampung people have always known what an earthquake is, so they make their houses earthquake-resistant. In the form of a house on stilts, the Lampung traditional house has stairs to access and enter the people who own the Lampung traditional house. Equipped with a small overhang called *anjungan,* which is always visible on the front or terrace of every house.

The pavilion serves as a place for the people of Lampung to have fun when there are no activities. Just hanging out with neighbours near the house.

A bee species record with a type locality in Central Sumatra (Rasmussen 2008)

Tetragonula fuscobalteata (Cameron, 1908)
Syn. *Trigona pygmaea* Friese 1933b: 147: Lectotype (ZMHB, worker): here designated "O-Sumatra / Mandau / 7.1933", "Trigona / pygmaea / Fr. / 1933 Friese det. / n."; paralectotype (DEI (1), ZMHB (1)) (common name (kloeloet itam ketjid)); **Type locality:** INDONESIA "Beringin, in der Wäldern oberhalb **Mandau (Sumatra), v. Bengkalis.** Im Juli 1933" (several workers);

---

[41] Sources: https://infolpg.com/rumah-adat-lampung/
Supplementary reading: https://rimbakita.com/rumah-adat-lampung/
https://www.gramedia.com/literasi/rumah-adat-lampung/

Sesat Balai Agung

Figure 130 An icon of Lampung's traditional house, Sesat Balai Agung

Becoming an icon of Lampung's traditional house[42], *Sesat Balai Agung* is a building whose main function is to meet traditional balancers or purwatin. There is a staircase called *Jambat Agung* when entering the Great Hall, also known as *Lorong Agung*. At the top of the stairs, there are three kinds of coloured umbrellas, namely white, yellow, and red. The colour is a symbol of the unity of the people of Lampung.

The white umbrella symbolizes the Lampung clan, the yellow umbrella symbolizes the social level in the village, and the red umbrella symbolizes the tribal level in the village. This traditional house also has a Garuda symbol on its roof because the animal is believed to be the vehicle of Lord Vishnu in ancient times. However, nowadays, the Garuda symbol is used as a seat for the bride and groom of Lampung traditional villagers.

Figure 131 Rumah Nuwou Balak (Rumah Kepala Suku)

Rumah *Nuwou Balak* (Rumah *Kepala Suku*)

Nuwou Balak, or the big house, is a residence for traditional balancers or tribal chiefs. With a size of 30×15 meters, the front of this traditional house is usually used to greet guests or relax. Uniquely, this traditional house has a separate kitchen from the main building. But it is connected to a building like a bridge. Divided into two meeting rooms, one for family meeting rooms, and eight bedrooms. Some of the rooms belonged to the wife of the head of the Lampung traditional tribe.

---

[42] https://en.wikipedia.org/wiki/Lampung

In front of the porch, some stairs connect to the ground or exit. Then, beside the bottom of the stairs is a *garang hadap,* the usual place to wash their feet before the Lampung tribe enters the house to maintain their cleanliness.

Figure 132 Rumah Adat Lampung Nuwou Lunik

Rumah Adat Lampung *Nuwou Lunik*

Switching from the house of the traditional tribal chief, *Nuwow Lunik* is intended for the ordinary people of Lampung village. The size is smaller than the *Nuwow Balak's* house belonging to the traditional tribal chief. It does not have a veranda and porch at the front, but a staircase near the entrance functions directly into the outer ground.

Having a smaller and simpler form than *Nuwow Balak*, this house only has a few bedrooms, and the kitchen is integrated into the main building. In the form of a roof, this house has a shape like an inverted boat or sometimes a pyramid.

The Meaning of the Parts of the Lampung Traditional House

Inside the house, several rooms have their functions in each room. Among others:
- *Pusiban* became the main and official place to hold deliberations between balancers.
- *Tetabuhan* is a place to store traditional musical instruments and Lampung traditional clothes.
- *Gajah Merem* is a place for the balancers to rest while doing the Traditional Pepung.
- *Kebik Tengah* is a place or container for children to balance sleep.
- *Anjungan* or *Serambi*, located outside the *pasiban* room, is usually this room to welcome guests of honour to small visits for *purwatin*.

43

Figure 133 Teluk Betung in the 1930s

43

https://en.wikipedia.org/wiki/Bandar_Lampung#/media/File:COLLECTIE_TROPENMUSEUM_Plein_met_kantoren_in_Te loekbetoeng_de_Lampongsche_Districten_op_Zuid-Sumatra._TMnr_60013127.jpg

Enggano Island, off Bengkulu Province, SW Sumatra

Kuano clan Traditional House - The **Enggano** people are isolated but in contact with the tribe that inhabits Enggano Island. A small island located adjacent to the southwest coast of Sumatra. The traditional architecture of Enggano Island until the early 20th century consisted of unusual round beehive-shaped houses. Source[44][45]

Traditional settlement Enggano settlements are structurally cumulus. Their houses have stilt frames, stacked and rectangular (whereas in the past, they were rounded), while the walls and rigid leaves strengthened the roof.

Figure 134 Kuano Clan Traditional House and a Model of a traditional beehive house

The Enggano people are one of the oldest tribes of Sumatra. The most anthropologically related people of the Enggano people are the Batak and Nias people and distantly related to the Lampung people (Abung and Pepaduan)

Rumah ulu[46] is a traditional house of people living upstream of the Musi River, **South Sumatra**. The name ulu is derived from the word *uluan*, which means "upstream". The term is also used as a generalization to rural inhabitants of the mountain range of the Central Bukit Barisan upstream of the river. The current province of South Sumatra encompasses only a

Figure 135 A sketch of the Rumah Ulu of the Uluan people of South Sumatra displayed in the Balaputradeva Museum.

small part of the former administrative region of South Sumatra (the present Sumbagsel[47] or Southern Region of Sumatra), consisting of the provinces of Bengkulu, Jambi, Lampung, and South Sumatra proper (the former Palembang Sultanate).

---

[44] https://en.wikipedia.org/wiki/Enggano_Island
[45] https://en.wikipedia.org/wiki/Traditional_architecture_of_Enggano
[46] https://www.wikiwand.com/en/Rumah_ulu
[47] **Sumatra Bagian Selatan** (Southern Region of Sumatra) includes Jambi, Bengkulu, SumSel – Sumatra Selatan (South Sumatra), the Bangka Belitung Islands and Lampung

Chapter 8

Javanese Architecture in Meliponiculture

By Abu Hassan Jalil

This topic is a favourite of mine. My motherland is Malaysia, but my mother's motherland is Java. She was of Betawi descent, the Batavian people of Jakarta, West Java.

Figure 136 AHJ family photo 1957

The author's late mother is adorned in the traditional Javanese *Kebaya* in this photo. This occasion was the celebration of Malaya Independence Day. In relating this to the Javanese architecture, we look at the Betawi traditional *Rumah Kebaya*.

The photo (left in Figure 126) is of one of the traditional houses in Indonesia, *Rumah Kebaya*, namely the traditional kebaya house from Jakarta. Source[48]

Figure 137 Left: Actual Betawi kebaya House; Middle: Model house to accommodate bee box; Right: Finished model built by Heri Damora in Lampung

Rumah Kebaya is a traditional house made of wood. The roof of the traditional kebaya house is saddle-shaped. The middle photo above is a replica of Betawi Rumah Kebaya bee housing. The photo on the left is the actual house, the middle is the miniature model, and the right is the completed varnished model. The video (https://www.youtube.com/watch?v=QW1oAF7G9Co) will show what elements were taken. Not all design elements were taken exactly; because of the aim to fill with the bees, we scaled the sizes for the top compartment and set up the miniature house with intentionally higher legs a little higher to keep out rainwater splashing while it is placed for decoration.

Betawi traditional house is the Kebaya House, also known as Bapang House. The roof shape can distinguish the kebaya and Joglo houses (Figure 129). The shape of the roof of the Kebaya house is

---

[48] https://en.wikipedia.org/wiki/Betawi_people

left and right, while the shape of the roof of a Joglo house is front and back, even though both have the same saddle pattern.

Unlike the Betawi Joglo or Stilt house, the Kebaya house always takes the form of a square or rectangular. The shape of the roof also has several pairs of roofs, making it look like a fold of kebaya (traditional blouse dress), eventually becoming the origin of the name Kebaya House. The attic is built like a horse's saddle and lined with an easel frame. The front of the house is provided with a basic roof called a sign or hat. This symbol works to withstand sun and rain exposure. The house is divided into two parts: the front and the inside. The front room is used for welcoming guests. While in the middle of the room was a family meeting place.

The Betawi Joglo house is square. Joglo house has three bedrooms: a front room, a living room and a back room. The front room is often called the front porch, and this room serves to accommodate guests. The living room is usually a dining room, a family gathering room and a bedroom. The influence of

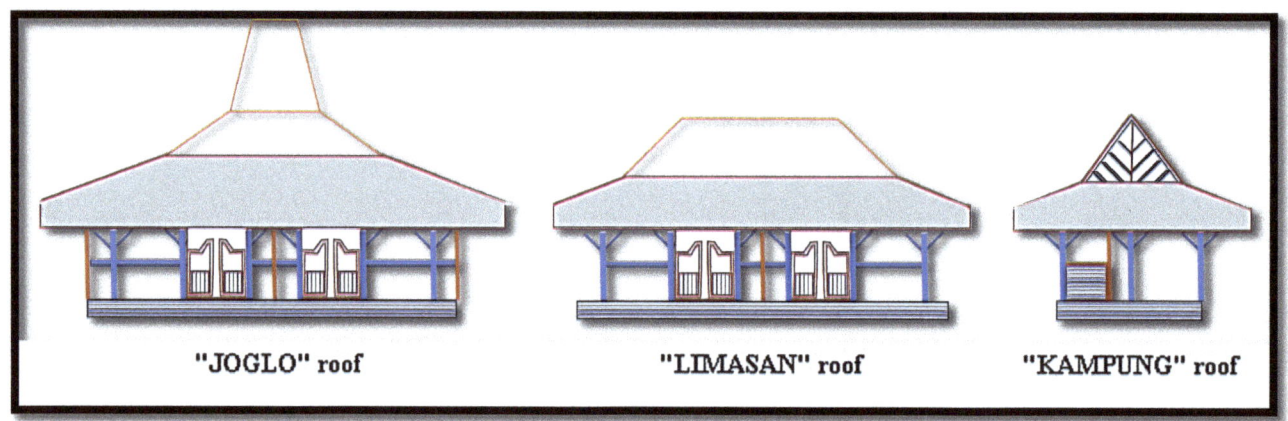

Figure 139 Examples of the Javanese Joglo, Limasan and Kampung roofs

Javanese culture is found in this rooftop house. After a closer look at the air movement, this data illustrates applying it below for Bee Housing, especially in the Bee Box Hive or 'Honey Super' roofing. Further reading: "Dragon and Bees of Java" ISBN 978-967-2290-13-1. Published 2018.

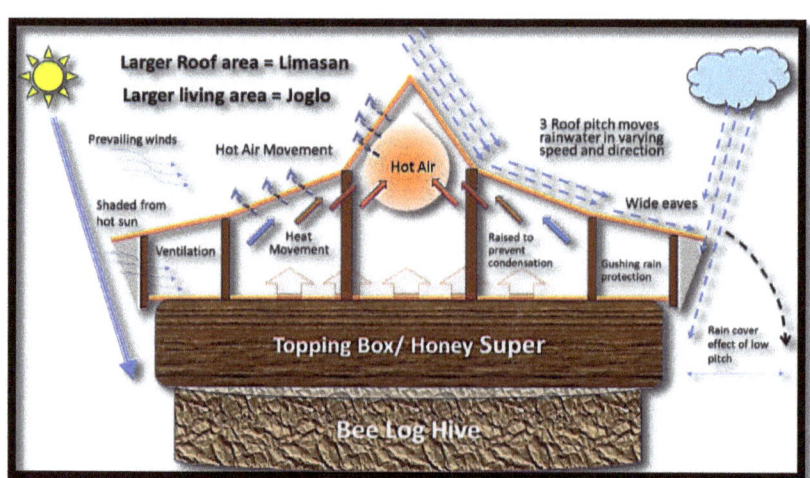

Figure 138 Illustration of weather protection of a Stingless bee topping box on a log hive with a Joglo-type roof.

The Javanese Joglo in other provinces of Java. These are some variations in Central Java.

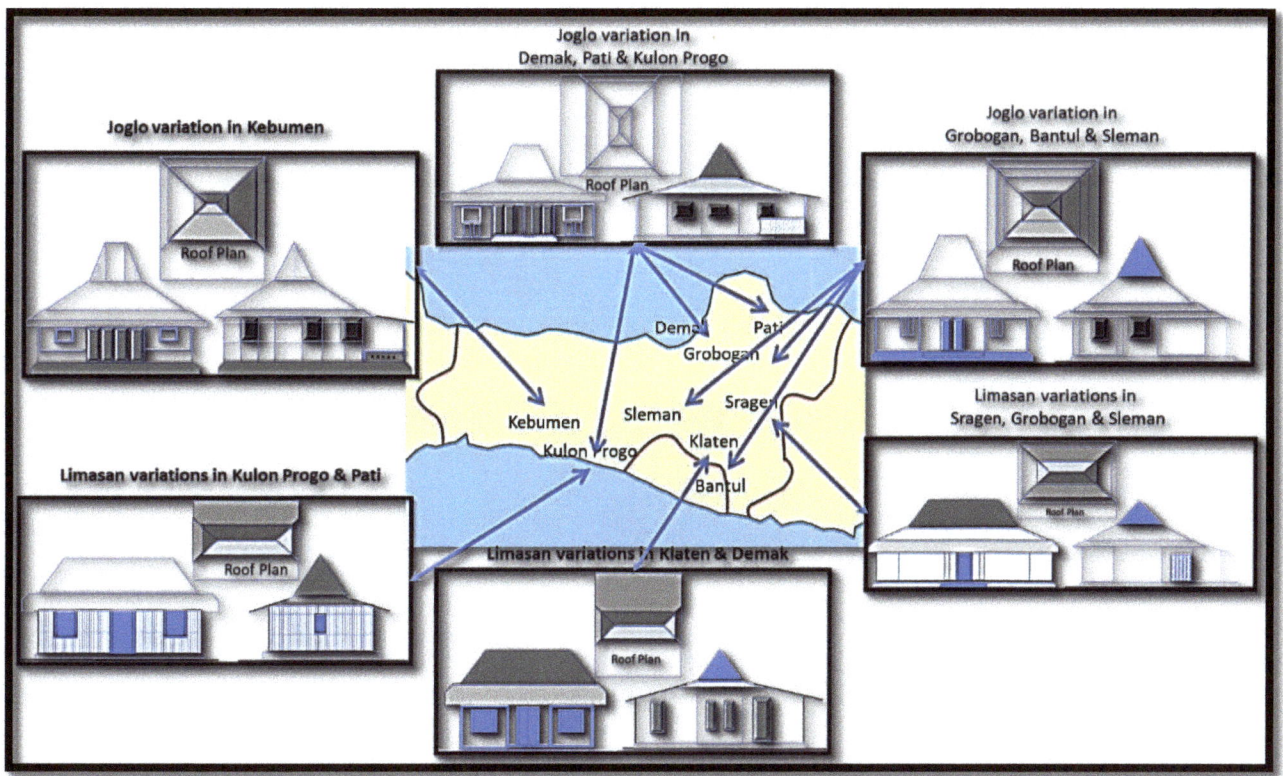

Figure 141 Various versions of Joglo and Limasan in Central Java

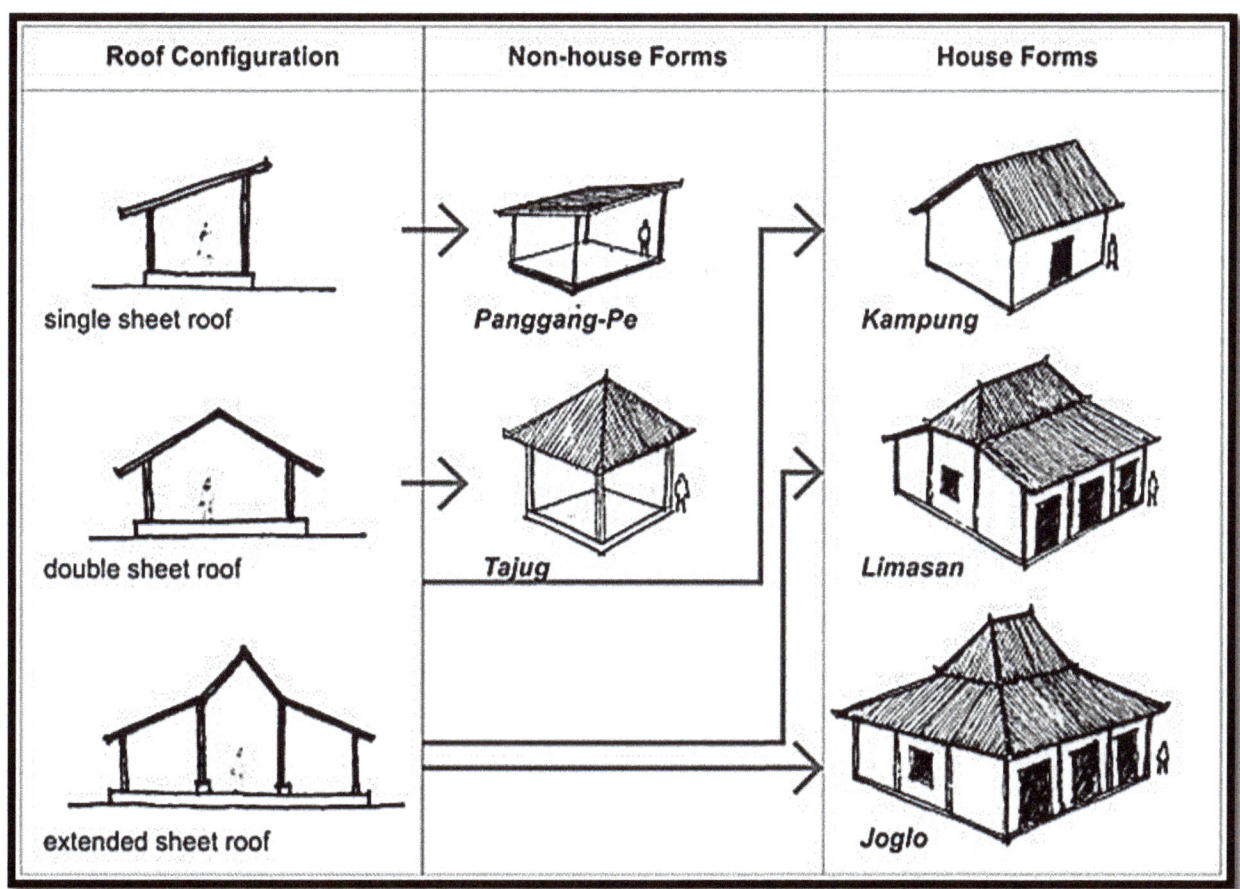

Figure 140 Roof configurations redrawn from anonymous public domain

Kampung roof is for common housing and is the simplest. It is a pitched gable roof erected over four central columns, braced by two layers of tie beams. The roof's ridge is supported by king posts and is aligned on a North-South axis. The roof can also be extended at a lesser inclination from the eaves of the existing roof.

Figure 143 The Javan Kampong roof is also found in South Sumatra. On the left is a model bee box hive roof built by Heri Damora.

- Higher-status Javanese families use *limasan* roofs. In Limasan houses, the basic ground plan of the Kampung style's four house posts is extended by adding a pair of posts at either gable end. A veranda extends the living space still further.

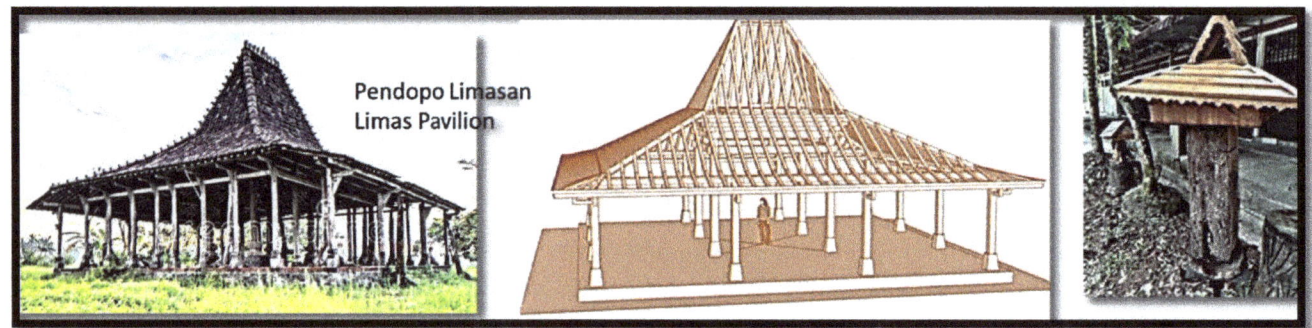

Figure 142 Limasan Pavilion (Pendopo in Java) type roofing.

- The Joglo roof is the most complex. It is associated with prestigious residences. The main roof is steeper, and the length of the four ridges is greatly reduced. Source[49]

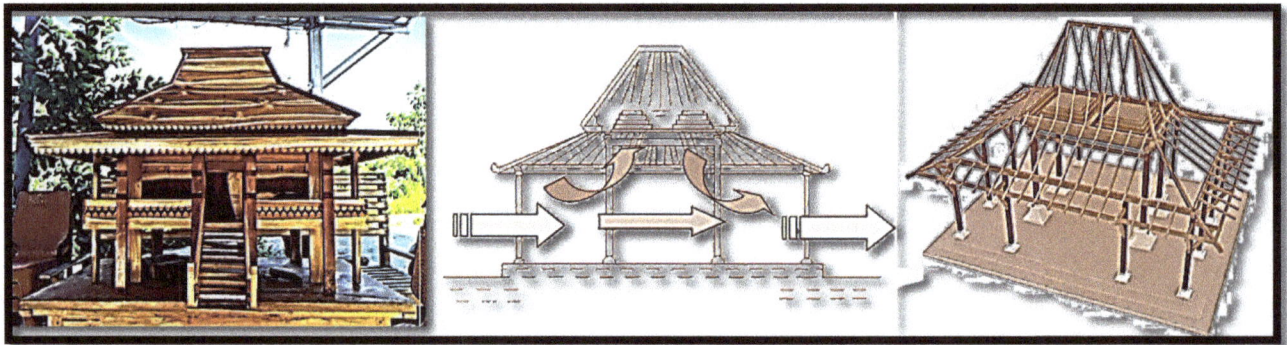

Figure 144 Javanese Joglo model by Heri Damora/ Middle photo shows the simulated air circulation and the right photo is the structure with rafter placement.

---

[49] https://en.wikipedia.org/wiki/Javanese_traditional_house

Bee Gallery - Prawita Garden, Banyumas, Java

A traditional house turned into a Bee Gallery.

Figure 146 The bee gallery in Prawita Garden in Banyumas, Java, visited by local tourists.

Visiting **Banyumas** some years ago in search of beekeepers having *Tetragonula drescheri,* the type locality for this species as described by the species' author. That's when we met with Mr. Teguh, the owner of Prawita Garden. He had a few colonies of *T. drescheri*, and while there, we were intrigued by their construction of a bee gallery. It was an extension of his traditionally built house. He has since extended his gallery to incorporate a café for visitors.

Figure 145 View of Prawita Garden Bee Gallery at the upper level.

A bee species record relevant to this locality (Rasmussen 2008)

Tetragonula drescheri (Schwarz, 1939)
Trigona (Tetragona) sarawakensis variety ***drescheri*** Schwarz 1939a: 85, 93, 106-107*: Holotype (AMNH); paratype (AMNH (1) (distribution, key to species, taxonomy); **Type locality**: INDONESIA "M. JAVA.- South **Banjoemas, Koebangkangkoeng,** 25 meters, July 1935 (F. C. Drescher)" (1 worker, holotype); "E. JAVA.-Mt. Goemitir, Nov. 3 (R. van der Veen)" (1 worker, paratype); Schwarz 1948: 118; Sakagami 1959: 120 (citation); Moure 1961: 210 (morphology, systematic position);

Sundanese Architecture of West Java

By Abu Hassan Jalil

At this juncture, we look back at Sundanese Culture. Native to the Sunda shelf, it has traditional Saddle roof architecture similar to the Batak Toba architecture. After scrutinising its structure (from the Batak Toba structures), we made this 'How to build a scissors clamp saddle roof' illustration for a Bee Box Hive.

Sundanese traditional house (Sundanese Imah adat Sunda) Sundanese traditional houses mostly take the basic form of gable-roofed structure, commonly called kampung style roof, made of thatched materials (*ijuk* or black Aren fibers, *hateup*[50] leaves or nipa palm leaves) covering wooden frames and beams, woven bamboo walls, and its structure is built on short stilts. Its roof variations might include hip and gablet roofs (a combination of gable and hip roofs). West Java traditional houses are called "Kasepuhan Houses."

Traditional Sundanesehouse forms include Buka Pongpok, Capit Gunting, Jubleg Nangkub, Badak Heuay, Tagog Anjing, and Perahu Kemureb. The ornamentation commonly includes the "o" or "x" shaped roof edges that are called capit gunting, which is very similar to a certain "x" design of Malay houses' roofs. The more elaborate overhanging gablet roof is called julang ngapak, which means "bird spreading wings".

Next to houses, a rice barn, called *leuit*[51] in Sundanese, is also an essential structure in the traditional Sundanese agricultural community. *Leuit* takes the basic form of a triangular gable-roofed structure made of thatched materials (ijuk black Aren fibers, hateup or kirai leaves, or palm leaves) covering wooden frames and beams, woven bamboo walls, and its structure is built on stilts, either short or long. Leuit is especially important during the Seren Taun harvest ceremony.

A more traditional house of Baduy people, a sub-ethnic of Sundanese people, is called *Sulah Nyanda*. It is commonly regarded as the blueprint of common Sundanese traditional houses. It is made from a wooden frame, woven bamboo wall,

The Badui or Baduy or Bedui Traditional House is a traditional house from Banten province. It is based on a traditional house owned by the Bedui. The Baduy / Badui tribe is an ethnic group of indigenous Banten tribes in the Lebak Regency, Banten. The form of a traditional Bedouin house is

---

[50] *Hateup* is the Sunda word for attap or thatch roofing material
[51] https://en.wikipedia.org/wiki/Leuit

often called the *Julang Ngapak*. The traditional Bedui house was built following the contour of the land where the building was built. The Bedouins are often said to be very environmentally sustainable.

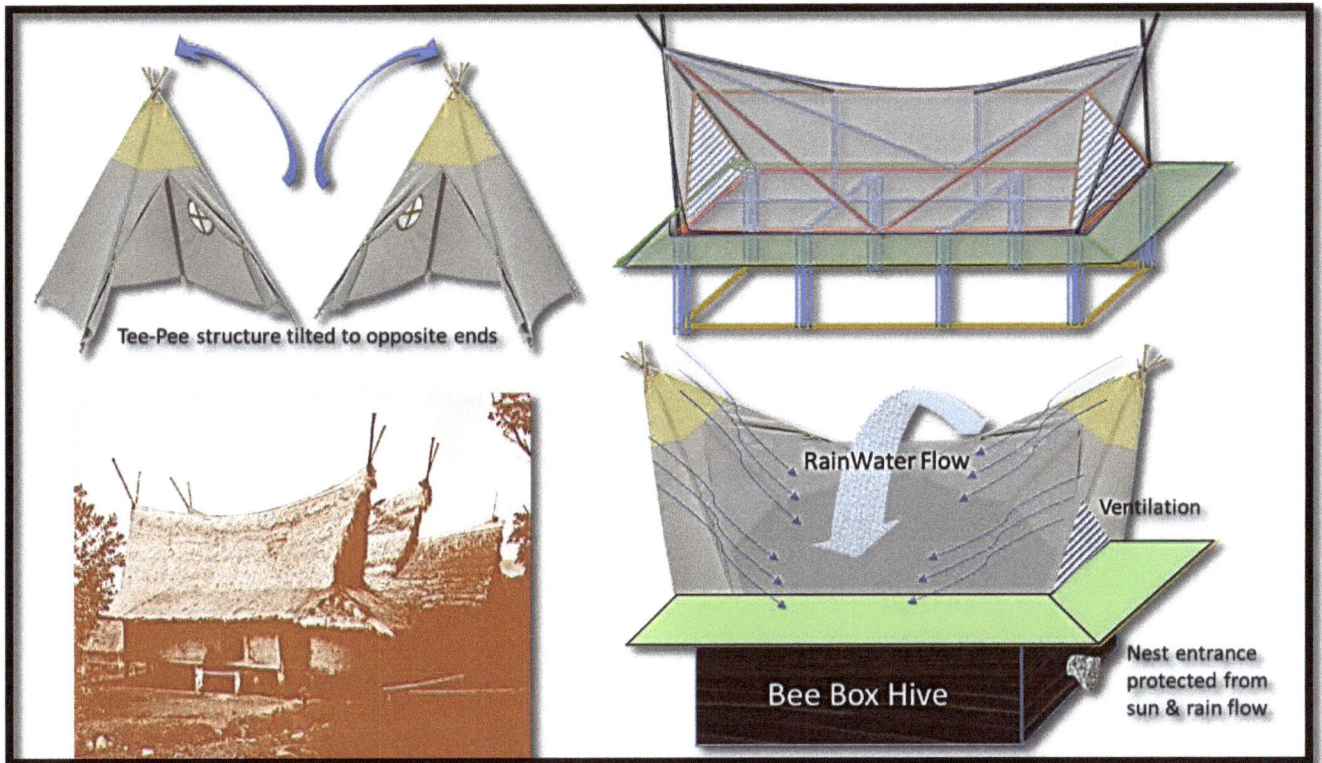

Figure 147 Assembling the Sundanese model with the saddle roof and scissors clamp finial/

Suppose we start with two tee-pee shapes and tilt them to opposite ends. No load-bearing trusses are necessary because this will be a miniature-scaled model. Add minimal beams and struts to keep them in place. The advantage of the saddle roof is that it keeps the rainwater flowing down the concave saddle so that the hive entrance can face the side and be protected. Prevailing winds may (if properly orientated) sweep hot air over the concave saddle and maintain thermal comfort for the bees inside.

On the right is a saddle roof similar to the Sundanese *Tagog Anjing*.

In this comparison, the Sundanese and the Betawi split to different ends. The earlier Sunda

Figure 148 Left: Betawi Joglo in Java; Right: West Javan Village hut model

saddle roof has a forked spire (*Capit Gunting* or scissors clamp) at both ends, while the Betawi of West Java adopts the Rumah Joglo or one type of Betawi Traditional House (Harun, 1991).

Rumah Adat or traditional house

The traditional house architecture of the Sundanese has never changed in terms of structure, although around the area inhabited by the Sundanese people, there are now magnificent buildings that show beauty as one of the results of modern architecture. Generally, a traditional Sundanese house is on stilts like other traditional houses in Indonesia. The form of this stilt house aims to avoid problems from the environment that can threaten its inhabitants. Judging by the shape of the roof, the traditional Sundanese house is divided into several characteristics that differ from one another:

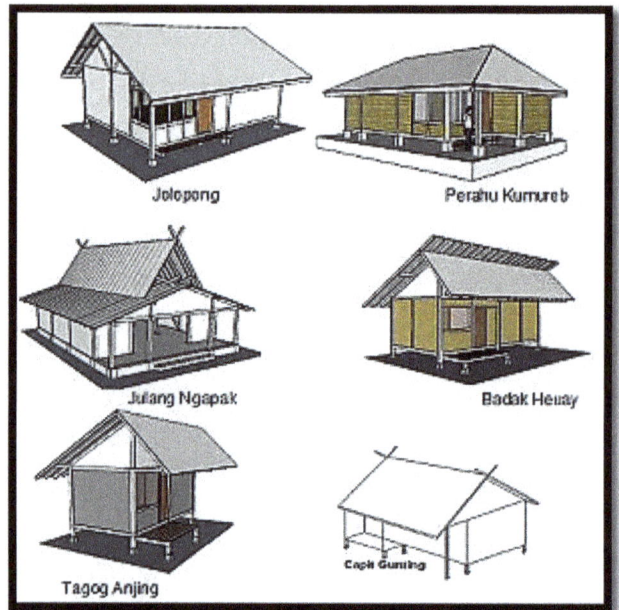

Figure 149 Rumah Adat or traditional house

Description:

1. *Jolopong* (a term for a house with an elongated gable roof)
2. *Kumureb* boat (the name for a house with a shield roof shape "by the Sundanese people, is called a *kumureb* boat because the roof shape is like an upturned boat").
3. *Julang Ngapak* (because the shape of the roof is like the wings of a bird in flight).
4. *Badak* (Rhinoceros) *Heuay* (because of the shape of the roof, like a rhinoceros yawning).
5. *Tagog Anjing* (Dog) (because the shape of the roof is like a sitting dog).
6. *Capit Gunting* Clamp Scissors (because the top of the roof is crisscrossed using scissors).

The Roof

The roof of the Sundanese house is made of palm fibre. The reason for choosing it as a roofing material is because it is a material that can absorb heat well, so it doesn't create a hot atmosphere in the house. The roof, as the crown of a building, has the function of protecting the occupants in it. The trellis on the front side of the house has a length of 2 meters. This makes the walls of the building not directly exposed to sunlight so that the walls as insulation are not hot and the space

Figure 1 Javanese dance in a backyard in Cirebon.

inside remains cold. In addition, there is also a side called the roof area made of woven bamboo, which serves as roof ventilation.

Location and Orientation

The traditional Sundanese house has a very neat layout and influences the community's belief that the house should not face the traditional earth (house). Thus, the orientation of the traditional Sundanese house always leads to the east and west.

This scrutiny is not simply on a self-indulged individual's house-building at their whim or fancy. These are traditional houses built with indigenous and heritage concepts based on cultural and traditional principles. Facing East is always a preference to get the mooring sun radiating through the front yard. It leaves evening activities in the front yard shaded by the building itself.

Kasepuhan House of West Java

The Banten Province Traditional House is known as a *Rumah Kasepuhan*. Here, the roof is a standard Gable roof, and it is the most applied shape by beekeepers to protect their box hives.

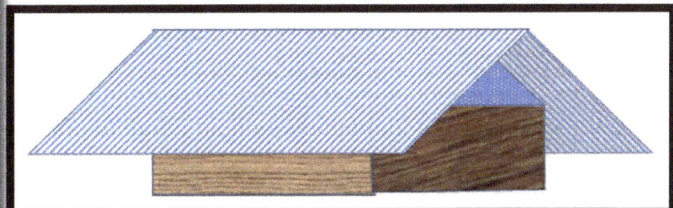

Figure 150 Banten Province Traditional House is known as a Rumah Kasepuhan. Right: Beekeepers' box hive roof

The West Java / Sunda Province Traditional House, also known as *Rumah Kesepuhan*, with a slightly different pronunciation. The roof shape here is a Hip and gable roof. This shape allows better rainwater runoff.

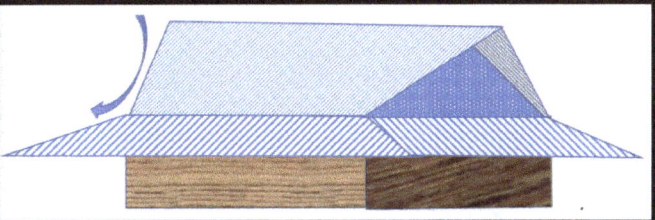

Figure 151 West Java / Sunda Province Traditional House is also known as Rumah Kesepuhan Right: Beekeepers' box hive roof

Central Java

The SB of Central Java is *Tetragonula* ***drescheri*** (Schwarz, 1939) [Typ. Loc.: Indonesia -M. Java. - South Banjoemas, Koebang Kangkoeng, July 1935 by Friedrich Carl Drescher] (Rasmussen, 2008). F. C. Drescher (1875-1957) was a plant collector in Malenesian Region. (Jstor, 2013).

The historical record of *T. drescheri* brought us to Banyumas in Java looking for it and ended up in a village Meliponary set-up beside the owner's house (Figure 143).

His house's double or overlapping gable roof was adopted on Bee box hives in Lampung, S. Sumatra. (See Penang Island House - *Rumah Gajah Menyusu* p.30). Here, we

Figure 153 A Beekeeper's home in Banyumas, Java.

Figure 153 Overlapping gable roof was adopted on Bee box hives in Lampung by Heri Damora

encountered several colonies of *T. drescheri* and got the opportunity to study them soon after we ventured into searching for the next iconic SB.

A bee record relevant to Java is:

Lepidotrigona javanica (Gribodo, 1891)
Trigona javanica Gribodo 1891: 109: Holotype (MSNG, worker). The terminata species group (taxonomy); Type locality: INDONESIA "Giava" (=Java) (1 worker); (Rasmussen 2008)

Figure 154 *Lepidotrigona sp specimen in the custody of Pak Teguh, Prawita Garden, Ajibarang, Banyumas*

Bawean Island, off Gresik Regency, East Java

Bawean (Indonesian: Pulau Bawean, also called Boyan in Malay) is an island of Indonesia located approximately 150 kilometres (93 miles) north of Surabaya in the Java Sea, off the coast of Java. The Gresik Regency of East Java province administers it. It is approximately 15 km (9.3 mi) in diameter and is circumnavigated by a single narrow road. Bawean is dominated by an extinct volcano at its centre that rises to 655 meters (2,149 feet) above sea level. The area is considered seismically active, with frequent tremors accompanied by landslides.

Figure 155 Model of a traditional Bawean home:

Image source[52]

The Island had little economic value and was used as a resting stop for ships sailing between Java and Borneo. Since the end of the 19th century, men of the island began to regularly travel to work in the British colonial possessions in the Malay Peninsula, especially in Singapore. Source

Figure 156 Banyunibo located in the center of paddy field southeast of Ratu Boko

Banyunibo[53] (Javanese: "dripping water") is a 9th-century Buddhist temple in Cepit hamlet[54], Bokoharjo village, Prambanan, Sleman Regency, Special Region of Yogyakarta, C. Java. The ancient temple dates from the era of the Mataram Kingdom.

Banyunibo has a curved rooftop design crowned with a solitary stupa; this theme is unique among the surviving Buddhist temples of Central Java. The curved rooftop was to symbolize lotus or *Padma* petals or mimic the organic roof made from ijuk fibres (black fibres surrounding the trunk of Arenga pinnata) common in ancient Java vernacular architecture and found today in Balinese temple roof architecture.

---

[52] https://en.wikipedia.org/wiki/Bawean#/media/File:COLLECTIE_TROPENMUSEUM_Model_van_een_huis_TMnr_H-534a.jpg
[53] https://en.wikipedia.org/wiki/Banyunibo
[54] a small settlement, generally one smaller than a village

Chapter 9

Traditional Architecture of Lesser Sunda and NTB-

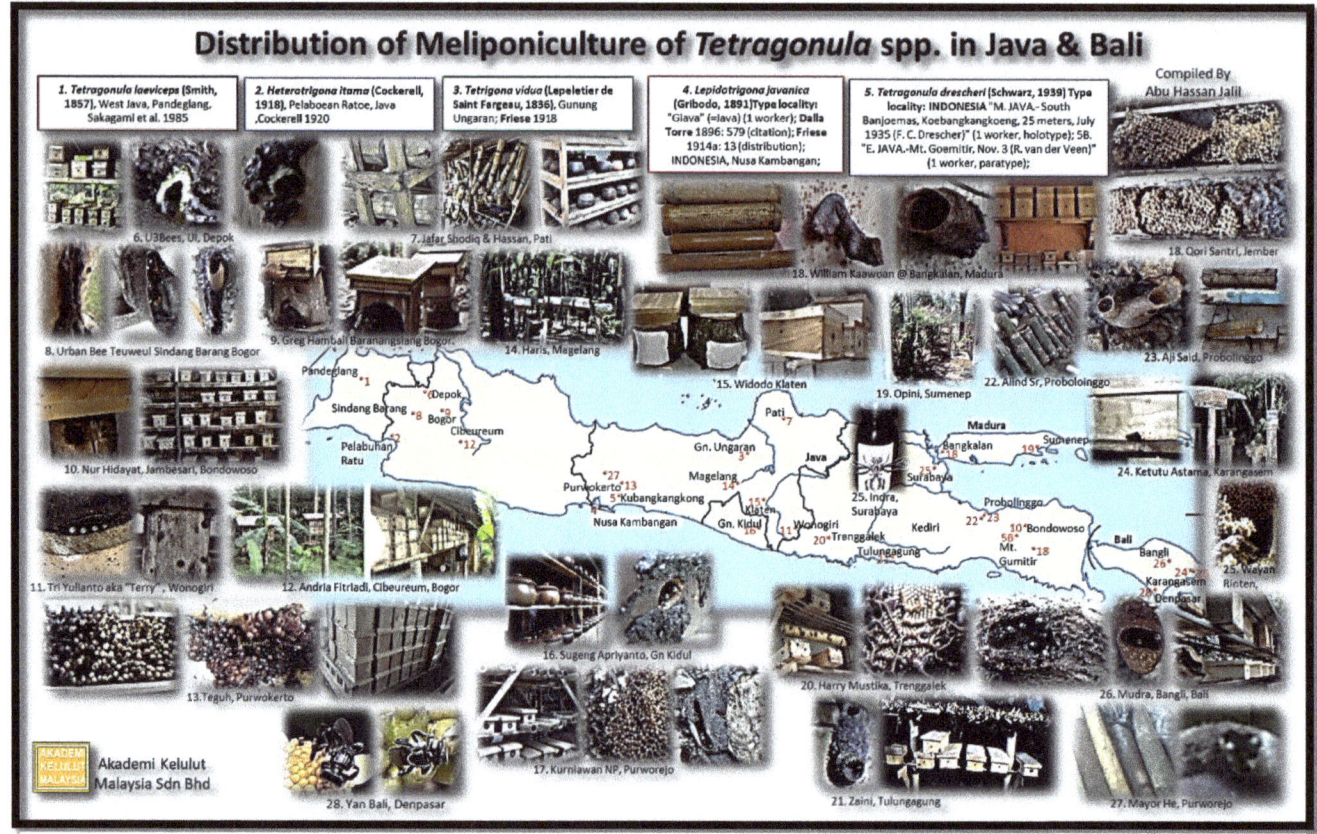

Figure 157 Distribution of Meliponiculture of Tetragonula spp. in Java & Bali

A study was done on the distribution of Meliponiculture activities in the Lesser Sunda Islands, which included Java and Bali. The operating procedures and most bee housing are similar, except that the roof of the hive boxes had some cultural influences and peculiarities in Bali.

Figure 158 Stingless Bee Farm in Bali

Balinese architecture

Next, we look at the ancestral cult of Bali. We visited Bangli, Bali, on a Meliponiculture tour with Dr. Bajaree Chuttong, Ms. Norita W. Pangestika (two fabulous upcoming ASEAN Bee scientists) and a couple from Jakarta (Apiary and Meliponary owners, Mr. & Mrs. Daniel Kustiono). We saw beehives in cuboid boxes and a log hive hung horizontally, reputed to be 43 years old and still teaming with stingless bees.

Figure 160 From Left Mrs. Mudra, Felicia Tjua Kustiono, Dr. Bajaree Chuttong and Ms. Norita W. Pangestika, Pak Mudra's mother, AHJ, Pak Mudra's father, Pak Mudra, and Daniel Kustiono.

We hoped to see some Balinese architecture adopted in the bee housing, but that was not the case. The Balinese beekeepers only maintain their heritage in their homes, ancestral shrines, and garden concepts. The Meliponary is a modern addition, although some have practised beekeeping for 50 years.

On the left below (Figure 147) is an ancestral shrine, and on the right photo is a garden lantern.

There are in no way the makings of a beehive. However, travelling to other regions is. I found other

Figure 159 On the left is an ancestral shrine and on the right pic is a garden lantern.

cultures that do replicate the Balinese design concepts.

I designed a Balinese-inspired Bee box hive rack structure (Figure 233). The bee housing roofs here were done in South Kalimantan (by a non-Balinese), replicating a Balinese Pavilion in Ubud, Bali. These incorporate the Balinese thatched roof done by non-Balinese, clearly aiming for Thermal Comfort while keeping the Indonesian identity using the "alang-alang" dried grass (like those of the Hawaiian grass skirts), also known as Cogon grass (Lalang in Malay -Imperata cylindrica).

Figure 162 The bee housing roofs here were done in South Kalimantan (by a non-Balinese), replicating a Balinese Pavilion in Ubud, Bali with a touch of oriental flair.

Figure 161 Examples of thatch roofing in Balinese styles done by beekeepers. Right: Traditional Balinese walled residential compound belonged to a common man.

Loloan Malays[55] or Balinese Malays (Malay: Melayu Loloan; Jawi: ملايو لولون) are a sub-ethnic group of the Malay who've lived in Loloan, Jembrana, Bali, Indonesia since the 17th century. There are approximately 28,000 up to 65,000 Loloan Malays living in Bali.

The Loloan Malays are predominantly Muslim, distinguished from the majority Balinese ethnic group, which is predominantly Hindu.

---

[55] Source: https://en.wikipedia.org/wiki/Loloan_Malays

Vernacular Architecture of Lombok Island

Etymology: The word Lombok is a word that comes from the local Sasak language. Translated into Indonesian, it means lupus or straight. Lombok is also a lesser-used word for chilli in Bahasa Indonesian, which has led many people to believe the island is named for its spicy cuisine.

On to Lombok Island to explore the Sasak Architecture. Another type found in Wallacea is the Gambrel roof, more likely Dutch colonial influence as a Barn roof, and in Lombok, the attic is used to store grain.

Figure 163 Gambrel roof, more likely Dutch colonial influence as a Barn roof, and in Lombok, the attic is used to store grain.

Gambrel roof - like a gable roof, but only if you add another slope to its lower edges. A gambrel's lower slope has a much steeper pitch, while the upper side is gentler. The simpler construction also allows the gambrel roof to use only two roof beams, and the Lumbung in Lombok brings the eaves down as low as possible to avoid strong wind uplift of the eaves. Gambrel roofs are often described as "barn roofs" of Georgian and Dutch Colonial influence.

Figure 164 An experimental Terrarium with a T. fuscobalteata hive in a branch hollow. A small miniature replica souvenir that can easily fit a tiny bee eduction. The 25cm x 13cm miniature version Lumbung for the meliponarium (Meliponine Terrarium).

Incidentally, I have used the miniature souvenir for the eduction of *Tetragonula fuscobalteata*.

The miniature souvenir was placed in a Terrarium with a small log nest of a *T. fuscobalteata* colony. Details of this are in a chapter in the Beescape book[56].

The Lumbung is built on stilts. It was known to withstand earthquakes when Mt. Rinjani was active. The traditional Sasak house has an attic to store grain. On the left is a replica Bee box hive.

A 25cm x 13cm miniature version was used to house a *T. fuscobalteata* colony for a meliponarium (Meliponine Terrarium). The experiment lasted for six months to explore the eduction of the colony in a controlled environment.

Figure 165 A Lumbung replica model bee box hive by Hasan Asri of Makassar.

Some of the regions in Lombok do not adapt to living in barn-type dwellings or grain stores. They have their indigenous ethnic architecture. Hasan Asri of Makassar, who also has a Beefarm in Lombok, built this bee box hive of the Lumbung design.

Figure 166 Traditional home of the Sasak tribe in Lombok

The Sasak of North (Tanjung District) and Central Lombok (Pujut District) built houses on split-level or Clerestory roofs. A traditional and simple home of a Sasak family in Lombok inspired this type of roof[57].

Stingless bees in Lombok favour the fruit stalks of the Arenga nut. The illustration below proposes a Bee house of a simple split roof structure with sufficient hanging hooks for the commonly dried gourd hives. This proposal shows that it is possible to keep together feral hives that form the stock for swarm traps. It incorporates an open cabinet as a transition store of the Arenga fruit stalk and branch hollow feral nests for swarm traps and colony eduction activities.

Figure 167 The oldest mosque dating from 1634 in Bayan.

---

[56] (https://books.google.com.my/books/about/Beescape_for_Meliponines.html?id=fUh9BAAAQBAJ&redir_esc=y)
[57] https://en.wikipedia.org/wiki/Lombok

Figure 171 Inspired by the Sasak Traditional house, N. and Central Lombok Is.

Figure 169 Rumah Budaya Majapahit, Trowulan, East Java

Figure 170 An old painting of a Majapahit house with an inset of a Majapahit shadow puppet scene.

Majapahit culture is a remnant of an ancient Hindu culture that still exists in a small village in Trowulan, East Java community. It was a Javanese Hindu-Buddhist thalassocratic empire in Southeast Asia that was based on the island of Java. It existed from 1293 to circa 1527.

The name Majapahit derives from Javanese, meaning "bitter maja". German orientalist Berthold Laufer suggested that maja came from the Javanese name of

Figure 168 Dried Maja fruits and gourds turn in kelulut hives.

Aegle marmelos, an Indonesian tree. The Maja fruit ripens at about 20 to 25 cm in diameter. With the innards removed, the outer rind dries to a hard globular shell.

Chapter 10

Meliponiculture Tourism & Ancestral Veneration in Ethnic Cultures of Sulawesi

For Torajans, the spirit of the dead is believed to linger in the world before the death ceremony is held. The Indonesian government has recognized this animistic belief as *Aluk To Dolo* ("Way of the Ancestors"). In 1984, the Indonesian Ministry of Tourism declared Tana Toraja Regency the prima donna of South Sulawesi. Tana Toraja is one of the conservation sites for the PROTO MALAY AUSTRONESIAN cultural civilization, which is still well-maintained today.

Figure 172 Celebrations in 1910-1940 miniature display houses comparing the image on the right of a full-scale house as tall as a coconut tree.

The display of miniature Toraja houses was a celebration from 1910 to 1940. These miniatures are compared to a full-scale house as tall as a coconut tree. South Sulawesi traditional houses like these are called "Tongkonan".

The traditional house is called *Tongkonan*. The word 'tongkonan' is derived from the Toraja word *tongkon* ('to sit'). Tongkonan is the centre of Torajan's social life. The rituals associated with the tongkonan are important expressions of Torajan spiritual life, and therefore, all family members are compelled to participate because symbolically, the tongkonan represents links to their ancestors and living and future kin.

The ethnic groups in the mountain regions of southwest and central Sulawesi (Celebes) are known by the name of Toraja, which has come to mean "those who live upstream" or "those who live in the mountains". Their name is derived from Raja, which in Sanskrit means "king". The society is hierarchically structured: the noblemen are called *rengnge*, the ordinary people *to makaka*, and the hierodule[58] *to kaunan*; their birth rite determines which rank a person will occupy.

The highly distinctive roofs constructed by the Toraja have given rise to various ingenious interpretations. The distinctive features of the Toraja's traditional houses (tongkonan) are the "buffalo

Figure 173 Depiction of air movement on a skeletal structure and a box hive model Toraja roof

horns", the roof design and the rich decoration on the walls. The buffalo symbolises status, courage, strength and fighting spirit. The tongkonan is constructed in three parts: the upper world (the roof), the world of humans (the middle of the building), and the underworld (the space under the floor) and designed as a representation of the universe,

Certainly, the roof is something of deep significance for the Toraja, and even today, they build "modern" (in other words, houses built with cement) houses with such roofs. These days, the miniatures are turned into stingless beehives. Novelty or bee thermal comfort? Something to ponder over the effort to put in for the bees.

While in Sulawesi, we shall explore other traditions in the different regions and provinces.

Gorontalo Province

Gorontalo[59] (Jawi: غارانتالي) is a province in Indonesia which is in the northern part of the island of Sulawesi. In the Dutch colonial era, the Gorontalo Province area was known as the "*Semenanjung Gorontalo*" (Gorontalo Peninsula) in the northern part of Sulawesi Island.

---

[58] (in ancient times) a slave living in a temple and dedicated to the service of a god.
[59] Source: https://id.wikipedia.org/wiki/Gorontalo

The location of Gorontalo Province is very strategic because it is flanked by two seas, namely the Gulf of Gorontalo, better known as the Tomini Bay, in the south and the Sulawesi Sea in the north. In the maritime history of the archipelago, the Sulawesi Sea is important because it is a shipping route from the island of Sulawesi to the Philippines, which also passes through the territorial waters of the Sulu Sultanate to the east of Malaysia.

The majority of the population in this area is the Gorontalo tribe, as well as being the tribe with the largest population in the northern peninsula of Sulawesi Island, followed by the Minahasa tribe in second place. The Gorontalo tribe is also a nomadic tribe with populations mostly found in North Sulawesi, Central Sulawesi, South Sulawesi, East Kalimantan, Java, and Papua.

The Origin of the Naming Gorontalo

According to historical saga records of traditional elders, the Gorontalo region was once a small island surrounded by the ocean. Over time, the seawater around the island receded and was followed by the emergence of three mountains to the surface. From several sources, the following is the origin of the naming of Gorontalo, which is mostly told from generation to generation, namely:

The word Gorontalo comes from *Hulontalangi*, which means Valley of Nobleness. Hulontalangi comes from two syllables, Huluntu, Valley, and Langi, Noble.

Derived from the word *Huidu Totolu* or *Goenong Tello*, which means "Three Mountains". If traced to its history, there are three ancient mountains on the Gorontalo peninsula, namely Mount Malenggalila, Mount Tilonggabila (changed to *Tilongkabila*) and another mountain that is not named. The valley south of Mount Tilongkabila is known as *Hulontalangi* or *Hulontalo,* which is also the forerunner of the Gorontalo City area. In much of the Portuguese and Dutch literature (and in some map drawings), the word Goenong-Tello is used more to describe this region.

Gorontalo comes from the word "*Pogulatalo*", which means "Waiting Place". The word "*Pogulatalo*" gradually changed in people's speech to "*Hulatalo*" or may come from the word "*Hulontalo*" or "*Holontalo*." The word Gorontalo is thought to be an adaptation of the name of the Hulontalo Kingdom in the past. The name Gorontalo itself was given by explorers such as the Portuguese and the Dutch who had come to Gorontalo.

Prehistoric Period

Figure 174 This bee house is inspired by the Dulohupa traditional house in Gorontalo, N. Sulawesi.

Based on research by the Archaeological Agency of Manado, North Sulawesi, it was found that Gorontalo has a prehistoric civilization site located in the southern region of Gorontalo. The research site was later named the "Oluhuta Site", a prehistoric site with an estimated age of more than 2000 years.

A bee species with a Type Locality in Sulawesi (Rasmussen 2008)

Wallacetrigona Engel and Rasmussen, new genus replacing *Geniotrigona incisa* (Sakagami & Inoue, 1989)

> (Previous combination): ***Trigona (Geniotrigona) incisa*** Sakagami & Inoue 1989: 605-610*, 614, 615, 617, 618, 619: Holotype (RMNH); paratypes (RMNH, six workers; MBBJ, one worker; SFS, remaining). Rasmussen & Cameron (2007) found that this taxon is distinct from *Geniotrigona,* and it is awaiting a new genus to be proposed[60] (taxonomy); **Type locality**: INDONESIA "Modoinding Minahasa, **N. Celebes**, vi. 26-27, 1941, native collector, ded. F. Dupont" (3 workers, holotype, paratypes); "Todyamboe, 900m, C. Celebes" (3 workers); "Wuasa, Kab. Poso, Sulteng, Sulawesi" (9 workers); "Lorei Lindu Nat. Park, C. Sulawesi" (3 workers); (Rasmussen 2008)

---

[60] RASMUSSEN, THOMAS, & ENGEL 2017; A New Genus of Eastern Hemisphere Stingless Bees (Hymenoptera: Apidae), with a Key to the Supraspecific Groups of Indomalayan and Australasian Meliponini, AMERIC AN MUSEUM NOVITATES Number 3888, 33 pp.

A new genus of stingless bees (Apinae, Meliponini) is described from Indonesia (Sulawesi), known from a single species previously placed in *Geniotrigona* Moure. Based on recent phylogenetic studies, *Trigona (Geniotrigona) incisa* Sakagami and Inoue renders Geniotrigona polyphyletic and is more closely related to *Lepidotrigona* Moure. The species is transferred to *Wallacetrigona* Engel and Rasmussen, a new genus, and differentiated from Geniotrigona proper and all other Meliponines occurring in Sundaland, Wallacea, and Sahul (Australinea). The new genus occurs east of the Wallace Line. It is separate from the distribution of Geniotrigona, which is otherwise restricted to Sundaland. Still, *Wallacetrigona* is presently not known beyond the Weber Line.

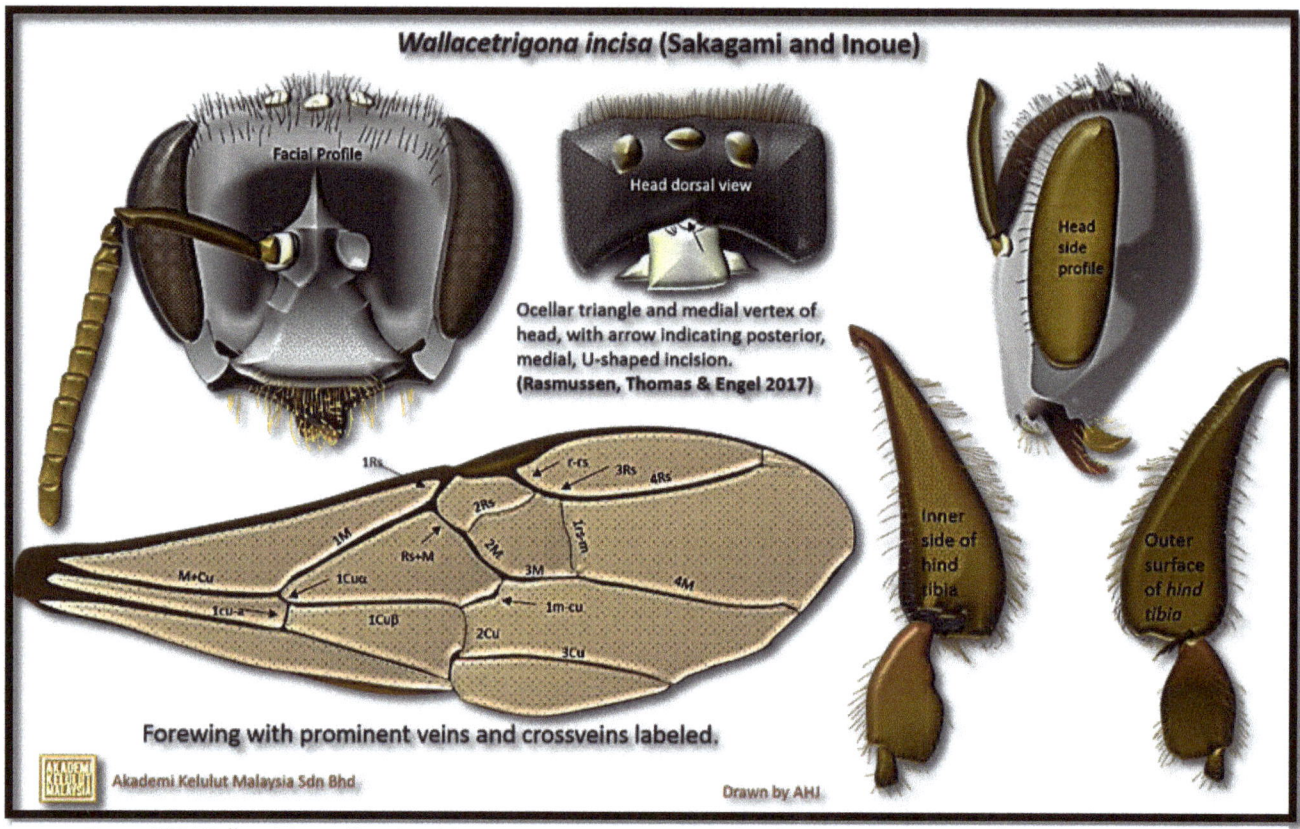

Figure 175 Wallacetrigona Engel and Rasmussen, new genus replacing Geniotrigona incisa (Sakagami & Inoue, 1989)

Vernacular architecture of Sulawesi

Figure 176 Vernacular architecture Sulawesi, Sumbawa, Maluku & Papua.00

1. *Rumah Pewaris,* or *Walewangko,* is a traditional house in the Minahasa area, North Sulawesi Province. Minahasa, formerly known as Tanah Malesung, was a peninsula where the Portuguese and Spanish settled.

2. *Dulohupa* is a traditional Indonesian house from Limba Village, Kota Selatan District, Gorontalo City, North Sulawesi Province. *Dulohupa* is a house on stilts with a body of boards and a roof structure with a Gorontalo nuance (Figure 159, p. 105). As a symbol of the Gorontalo traditional house, *Dulohupa* has wooden pillars, while as a symbol of the traditional ladder or what is also known as *Tolitihu*, *Dulo* in an appendix element *Bhupa* (combining the two terms to *Dulohupa*) has two stairs, each on the right and left of the house.

Figure 177 Menado Traditional House: Rumah Pewaris

3. *Boyang* House is a traditional house from West Sulawesi Province. *Rumah Boyang* has a unique architectural style, consisting of a house on stilts composed of wood and supported by supporting

poles. This house is the residence of the Mandar Tribe, which is an indigenous tribe from West Sulawesi. The pediment and Gable end is like the Banua Layuk (see Figure 164)

4. *Rumah Tambi* is a traditional house or traditional house from the province of Central Sulawesi, Indonesia. This traditional house is in the form of a stage whose roof is also useful as a wall. *Rumah Tambi* is the home of the Kaili and Lore tribes, which are generally the homes of residents, and several areas in Central Sulawesi make this house the home of traditional chiefs.

5. Southeast Sulawesi. The population of Southeast Sulawesi consists of several tribes, namely Tolaki, Buton (Wolio), Muna, Mekongga, and Kaba Ena. On the island of Kaba Ena, the people are generally Muslim. Residential houses in the Tolaki tribe are called *Laika (Kon awe)* and *Raha (Mekongga)*, which means house. *Laika's* house is a residence of the Tolaki[61] tribe.

Figure 178 Laika House, Kendari, SE Sulawesi

Figure 179 Banua Layuk Traditional House of West Sulawesi / Mamuju Regency.

6. *Banua Layuk* Traditional House of West Sulawesi / Mamuju Regency.

Mamuju Regency is a regency (Indonesian: kabupaten Mamuju) of West Sulawesi province, Indonesia. The regency capital is Karema, while Mamuju town is the capital of West Sulawesi.

7. *Banua Sura*[62] Mamasa Traditional House. The house, commonly called Banua, is also a characteristic of the Mamasa community. The typical Mamasa house has similarities to the traditional Toraja house, which is now increasingly rare, and the types are also different. One is the *banua sura*, or carved

Figure 180 Banua Sura Rumah Adar Mamasa

house for the nobles. Located in the interior of the Mamasa capital, namely in Waka Hamlet,

---

[61] https://media.neliti.com/media/publications/92494-ID-bentuk-fungsi-dan-makna-interior-rumah-a.pdf
[62] https://id.m.wikipedia.org/wiki/Berkas:Rumah_Adat_Mamasa_(Banua_Sura).jpg

Salutambun Village, Pana District, Mamasa Regency, West Sulawesi Province, it still stands strong, but the roof has been replaced, even the age of this traditional house is hundreds of years. (Source[63]:)

8. The Bugis[64] people, also known as **Buginese**, are an ethnicity—the most numerous of the three major linguistic and ethnic groups of South Sulawesi in the southwestern province of Sulawesi, the third-largest island of Indonesia. The main religion embraced by the Bugis in 1605 is Islam, with a small minority adhering to Christianity or a pre-Islamic indigenous Animism belief called *Tolotang*.

Figure 181 A traditional Bugis dwelling in South Sulawesi

9. The Sangihe Islands[65] (also spelt "Sangir", "Sanghir", or "Sangi") – Indonesian: Kepulauan Sangihe – are a group of islands that constitute two regencies within the province of North Sulawesi, in northern Indonesia, the Sangihe Islands Regency (Kabupaten Kepulauan Sangihe) and the Sitaro Islands Regency (Kabupaten Siau Tagulandang Biaro). They are located northeast of Sulawesi between the Celebes Sea and the Molucca Sea, roughly halfway between Sulawesi and Mindanao, in the Philippines; the Sangihes form the eastern limit of the Celebes Sea. The islands combine 813 square kilometres (314 sq mi), with many islands actively volcanic with fertile soil and mountains.

Figure 182 "Tarempa" en kantoor, Poelau Sangihe

10. **Buton** (also Butung, Boeton or Button) is an island in Indonesia located off the southeast peninsula of Sulawesi. In the precolonial era, the island, then usually known as Butung, was within the sphere of influence of Ternate in Maluku.

Figure 183 Buton, Sulawesi (region)

---

[63] https://en.wikipedia.org/wiki/Mamasa_Regency
[64] https://en.wikipedia.org/wiki/Bugis#/media/File:Bugis_house.JPG
[65] Source: https://en.wikipedia.org/wiki/Sangihe_Islands

Chapter 11

Vernacular architecture in Kalimantan.

Etymology. The name Kalimantan is derived from the Sanskrit word Kalamanthana, which means "burning weather island", or island with a very hot temperature, referring to its hot and humid tropical climate.

On January 18, 2022, Indonesia's Parliament passed the Capital City Bill into law, meaning constructing the country's new capital can begin ahead of the planned relocation from Jakarta starting in early 2024 to East Kalimantan province.

This chart of Borneo is for anyone who cares enough to set up a project to house models of these vernacular miniatures to house bees because many of them (both the bees and the rare traditional structures - as some listed below) may start fading away by 2024.

1. ***Rumah Bubungan Tinggi*** or *Rumah Banjar* or *Rumah Ba-anjung* is the most iconic type of house in South Kalimantan. In the old kingdom period, this house was the core building in a palace complex. This house is where the King and his family would reside. Since 1850, various buildings have been around it with their respective functions.

   Figure 184 Banjar Bubungan Tinggi House

   "*Bubungan Tinggi*" refers to its high pitch roof (45 degrees inclination). This type of house became so popular that people out of the royalty also took an interest in building it. Hence, houses with this type of architecture are all over South Kalimantan, crossing Central Kalimantan's and East Kalimantan's borders. This type of house, of course, took more money than the usual house, so it was naturally the house of the rich.

2. **The Baloy[66] Traditional House** is a typical traditional house of North Kalimantan. This traditional house is in the form of a stage built using ironwood because it is very hard. Ironwood is also used as a building material for other traditional houses in Kalimantan.

---

[66] https://www.rumah.com/panduan-properti/rumah-adat-kalimantan-utara-60111

This traditional house has a characteristic form of carvings with coastal motifs, mostly occupied by the Tidung[67] tribe. The direction of the Baloy house is unique in that it is arranged in such a way that it faces north, with the main door facing south.

Figure 185 Baloy Traditional House of North Kalimantan Source:

3. **Lamin houses** are usually decorated with certain motifs consisting of carvings, sculptures, and picture motifs. For the Dayak tribe, the decorative motifs of *Lamin* House,

Figure 186 Lamin House of East Kalimnatan Dayak

including carving, motifs, and paintings, are called *Lamin* ornaments. The ornaments or accessories are in the form of a decorative motif that has a circular pattern. The typical colours of *Rumah Lamin*[68] are yellow and black. In addition, several other colours are often used in *Lamin* ornaments: white, blue, and red. For the Dayak people, the colour yellow means authority. In addition to yellow and black, *Rumah Lamin* also uses a lot of blue, red, and white colours. The blue colour means loyalty, the red colour means courage, and the white colour means cleanliness of the soul. Black colour is usually used to colour the base of the wall.

4. **Rumah Undur Undur** in Ketapang, West Kalimantan. Ketapang is the local name for the Indian almond tree. A beekeeper would put kelulut colonies around the house and hang them wherever possible.

The ketapang (*Terminalia catappa*) tree by the house has just grown.

Figure 187 Rumah Undur Undur, Ketapang, Kalimantan

---

[67] https://id.wikipedia.org/wiki/Rumah_Baloy#/media/Berkas:Baloy_Mayo_Adat_Tidung_(2).JPG
[68] https://commons.wikimedia.org/wiki/File:Lamin_Adat_Pemung_Tawai_Samarinda.jpg

5. **Rumah Betang**, the Kalimantan traditional house, is found in various parts of Kalimantan and is inhabited by the Dayak community, especially in the upstream areas of the river, which are usually the centres of Dayak tribe settlements[69]. The shape varies among different Dayak communities in this region.

Figure 190 Rumah Betang, a traditional Ma'anyan house in Muara Bagok, East Barito Regency, Central Kalimantan.

Figure 189 Rumah Betang, Dayak Ngaju Longhouse, Central Kalimantan *Figure 189 Tapered Bee box hive topping a log hive.*

---

[69] https://en.wikipedia.org/wiki/Ma%27anyan_people

6. The name **Bidayuh** means 'inhabitants of the land'. Bidayuh is the collective name for several indigenous groups found in southern Sarawak, Malaysia and northern West Kalimantan, Indonesia, on the island of Borneo, which are broadly similar in language and culture (see also issues below). Originally from the western part of Borneo, the collective name Land Dayak was first used by Rajah James Brooke, the White Rajah of Sarawak. At times, they were also lesser referred to as Klemantan people.

Figure 191 A traditional Bidayuh 'baruk' roundhouse in Sarawak, Malaysia. It is a place for community gatherings. Source: https://en.wikipedia.org/wiki/Bidayuh

This Bidayuh roundhouse is very similar to the Melanau tribe house in Sarawak.

7. **Sandung**[70] or sandong is the ossuary of the Katingan, Ngaju and Pesaguan people native to Indonesia's southern and central Kalimantan. The sandung is an integral part of the Tiwah ceremony of the Ngaju people, a secondary burial ritual where the bones of the deceased are taken from the cemeteries, purified, and finally placed in a sandung.

Figure 192 Left: A sandung in Central Kalimantan ca. 1915.; Right: Details of a sandung of Pesaguan people in Ketapang Regency, West Kalimantan. Note a sculpture of a dragon above it.

List of records of Stingless bees found in Borneo (Rasmussen 2008)

Trigona borneënsis Friese 1933a: 46: Lectotype (ZMHB, worker): here designated, "Borneo / Sanggau / 24-7-32", "*Trigona / borneensis* / 1925 Friese det. Fr."; paralectotypes (DEI (4 (2 on one pin), ZMHB (2)); possible additional types (AMNH, USNM) (taxonomy); **Type locality**: INDONESIA "Sanggau (**Borneo**) 24. July 1932" (several workers);

Record of a synonym to ***Tetrigona apicalis*** (Smith, 1857)

Trigona sericea Friese 1933a: 45-46: Lectotype (ZMHB, worker): here designated, "Borneo / Sanggau / 24-7-32", "Trigona / sericea / 1925 Friese det. / Fr."; paralectotypes (ZMHB (5), DEI (4)) (taxonomy); Type locality: MALAYSIA/INDONESIA "Meliau (Borneo) 4. June 1932"; "Sanggau (Borneo) 24. Juli 1932"(several workers);

Record of ***Tetragonula melanocephala*** (Gribodo, 1893)

---

[70] https://en.wikipedia.org/wiki/Sandung

Trigona melanocephala Gribodo 1893: 264: Holotype (MSNG, worker): The authentic holotype of this taxon is labelled "Bandjarmas" (?=Bandjarmasin, near Liangtelan) (F. Penati, pers. com.) (taxonomy); Type locality: MALAYSIA "Liangtelan (Borneo)" (1 worker); Dalla Torre 1896: 580 (citation); Cockerell 1919b: 242 (distribution); MALAYSIA, Sandakan;

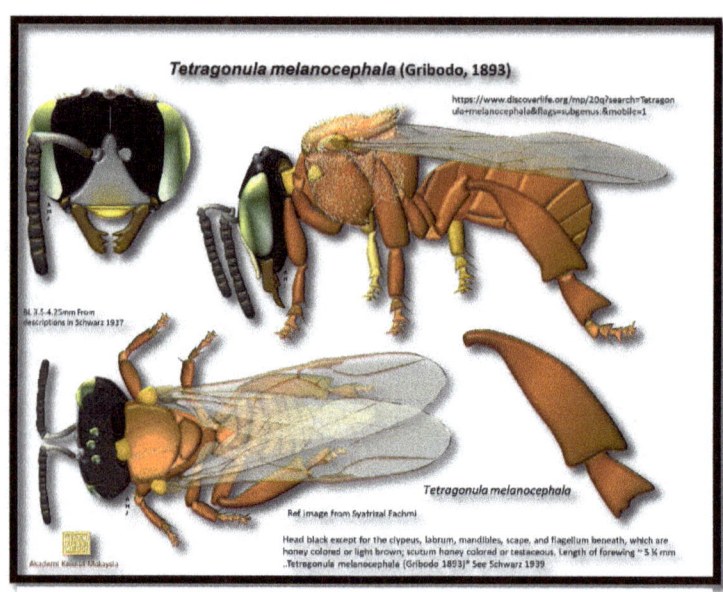

Figure 193 Trigona melanocephala Gribodo 1893

Kalimantan Vernacular Architecture Chart

Figure 194 Vernacular architecture in Borneo, Java, Bali & Lombok Island.

I ventured down to South Kalimantan to explore some indigenous domicile designs. I met with Ozie (real name Fauzi) in Barabai, west of Banjarmasin City. Posing for a photo. Ozie and his family are dressed in Banjar Dayak traditional costume in front of his designed Dayak house. The middle photo

Figure 196 Left: Meliponary in Barabai west of Banjarmasin city, South Kalimantan; Middle: A totem in front of Ozie's house. Right: Ozie and his family are dressed in Banjar Dayak traditional costume in front of his Dayak house design.

is a Banjar Dayak Totem between his house and the bee garden. Initially, he built blue bee housing with Balinese-inspired architecture, and as his clientele grew from mostly Singapore Chinese, he changed to oriental red.

Figure 195 Initially, Ozie built blue bee housing with Balinese-inspired architecture, and as his clientele grew from mostly Singapore Chinese, he changed to oriental red.

East of Bali along the Indonesia Archipelago is the Nusa Tenggara Barat, i.e., the West Nusa Tenggara group of Islands, including Lombok Island (see Chapter 8) and Sumbawa. Beyond that is the East Nusa Tenggara.

Chapter 12

Meliponiculture in Indonesian Wallacea (NTT region)

Vernacular architecture of Sumbawa

Figure 197 Traditionally Sumbawa Island

Etymology: Sumbawa is a Portuguese corruption of the locally used name Sambawa (still found as such in Makassarese, cf. also Semawa in the Sumbawa language). This name is probably derived from Sanskrit śāmbhawa, meaning 'related to Śambhu (= 'the Benevolent', a name for Shiva)'.

1. *Rumah Adat Dalam Loka* is the residence of kings from Sumbawa Regency, West Nusa Tenggara. This house in Loka, or Sumbawa palace, is a historical legacy of the Sumbawa kingdom. *Istana Dalam Loka* was built in 1885 by Sultan Muhammad Jalalludin III (1883-1931). The traditional house in Loka is the original design of the residence of the kings of Sumbawa. The strong influence of Islamic culture that entered this area at that time had made almost all aspects of the adat and ethnicity of the Sumbawa people dissolve in Islamic Sharia values.

2. *Musalaki* houses[71] are traditional houses often found in East Nusa Tenggara, Indonesia. This house itself is a symbol of the province of East Nusa Tenggara. This traditional house is a special residence for the chiefs of several tribes in the province of East Nusa Tenggara. Because it has become a symbol of the province, currently, the design of government buildings such as districts, sub-districts, and regencies in East Nusa Tenggara mostly adopt the concept of the Musalaki house, and in some areas, this house is already inhabited by the general public. It is a Hip roof but with an extended tall ridge. (See also Musalaki in Part 2 of Volume 3)

Figure 199 Rumah Adat Musalaki NTT *Figure 198 Kupang Traditional House: Sao ata mosa lakitana (Musalaki) House*

---

[71] Source: https://pariwisataindonesia.id/budaya-dan-sejarah/pariwisata-indonesia-rumah-musalaki-tertarik-untuk-mengenal-rumah-adat-dari-provinsi-ntt/

3. Rumah Bima / Suku Dompu[72]

    Uma Lengge[73] comes from the word *uma*, which means house and *lengge*, meaning conical. This traditional house is part of the cultural heritage of the Bima ancestors. The main material comes from wood, four pillars, and a roof formed from reeds. This traditional building consists of three floors. Receiving guests and performing traditional ceremonies

Figure 200 Uma Lengge - Rumah Bima / Suku Dompu

are performed on the first floor, the bed and kitchen are on the second floor, and the last floor is used as food storage or granary.

As times progressed, the people of Bima no longer used *Uma Lengge* as their original function. Many people choose to live in a larger conventional home. Nevertheless, the traditional house building is maintained; only its function is now fully used as a rice granary.

In Bima Regency, most people living in villages still use stilt houses to live. Bima vernacular architecture typology is *Uma Lengge, Uma Mbolo, Uma Jompa,* and *Uma Panggu* (Figure 179). *Uma lengge* and *Uma mbolo* still exist with a minimal amount. Meanwhile, the granaries called *Uma jompa* only survive in a few villages.

The traditional houses of the Dompu[74] tribe

Figure 201 Left: Uma Mbolo; Middle: Uma Jompa & Right: Uma

are named *Uma Jompa* and *Uma Panggu*. *Uma jompa* functions as a place to store rice granaries. It is located separately from the residence of the Dompu tribe. *Uma jompa* has three floors and functions much like the three-storey Bima houses. *Uma panggu/Uma ceko* is a house and a place to live for the Dompu people. This building is made of wood in the form of a stage or panggung in Malay.

---

[72] https://steemit.com/esteem/@ekafajarna/uma-lengge-house-the-last-bima-food-defense-6d1431df2d54f
[73] Hariyanto et al. A Simple Stilt Structure Technique for Earthquake Resistance of Wooden Vernacular Houses in Bima, Sumbawa Island, Indonesia *International Journal on Advanced Science Engineering, and Information Technology* Vol.12 (2022) No. 4 ISSN: 2088-5334
[74] Source: https://id.wikipedia.org/wiki/Suku_Dompu

Vernacular architecture of Maluku Islands

Figure 202 Maluku / Ambon Traditional House: Baileo

Etymology: Maluku can come from "Moloku", here meaning to grasp or hold. In this context, "Moloku Kie Raha" means "confederation of four mountains". However, the root word "loku" comes from the local Malay Creole word for a unit, therefore not from an indigenous language.

1. Maluku / Ambon Traditional House: *Baileo*[75]

    Baileo House (Figure 131). This traditional house is the most found and widely used, representing the Maluku traditional house. The *baileo* traditional house belongs to the Huaulu tribe, which is a native of Seram Island, Ambon.

    As a characteristic of this Maluku traditional house, the baileo house is shaped like a stilt house with a higher floor with no walls and windows. It has a square shape with a foundation made of wood, boards, and sago or sago leaves as the roof.

    At the front of the entrance, there will be a ladder measuring about 1.5 meters. Then, around the house, especially on the wooden supports, there will be carvings of chickens or dogs in pairs, colourful moons, stars, and suns.

    Another uniqueness can be found in the base of the baileo house. This traditional Maluku house has a pedestal with boards only arranged on the frame without nails. Although not nailed, the floor of the Baileo traditional house does not shift or creak because of the one who built it using a lock technique.

A valid bee name associated with the locality: ***Platytrigona keyensis*** (Friese, 1901), (Rasmussen 2008)
   Trigona keyensis Friese 1901: 271: Lectotype (ZMHB, worker): here designated, "AsiaArch. / Key Ins. / 1900 / Kühn", "Trigona / keyensis / 1900 Friese det. / Fr.", "Type" (red label), "Coll. / Friese"; paralectotype (AMNH (1)). The planifrons species group (manuscript name (latigena in litt.), taxonomy); **Type locality**: INDONESIA "**Key-Eilanden**[76] (**Amboina**, Nederland India) Durch H. Kühn" (6 workers);

---

[75] Source: https://www.orami.co.id/magazine/rumah-adat-maluku
[76] https://en.wikipedia.org/wiki/Kai_Islands (a group of islands in the Southeastern part of the Maluku Islands)

2. *Sasadu*[77] is the traditional house of the Sahu tribe in West Halmahera, also the original and oldest ethnic group in the area. In this house, the indigenous Sahu people usually gather in meetings. In West Halmahera, this house is common in every village. The use of Sasadu as a

Figure 203 Inspired by the 8-sided Pavilion of the Sasadu people on N. Maluku.

community meeting location is usually associated with holding various events, such as rituals or traditional ceremonies such as harvest celebrations and the election of traditional leaders, and welcoming guests who come. However, Sasadu can also be used to relax without special events. Etymologically, Sasadu comes from the word *sadu,* which in the Sahu language does not have any meaning, while in Ternate, it means to draw, and *sado* means complete.

3. *Rumah Hibualamo* Although it looks more modern, this traditional Maluku house is the oldest type of house in Maluku, believed to have existed for hundreds of years. The naming of the hibualamo house is taken from the words *'hibua',* which means house and *'lamo',* which means big. So that means a big

Figure 204 Rumah Adat Maluku Hibualamo (photo credit: abdulromli.it.student.pens.ac.id)

---

[77] Source: https://id.wikipedia.org/wiki/Sasadu

house. Unlike the previous two traditional Maluku houses, the hibualamo house has walls like modern houses.

Characteristic:

As explained above, the Hibualamo house has a more modern form because it has walls made of brick and cement. But let's look at the roof of this traditional Maluku house. The shape is very traditional because it resembles other Maluku traditional houses resembling a boat. The colours used in the Hibualamo House must be five colours, namely red, black, gold, yellow, and white.

The construction of a traditional Maluku house whose roof resembles a boat reflects the life of the Tabo tribe, most of whom are sailors and fishermen. In contrast, the octagonal shape and the four entrances represent the symbols of the four cardinal directions. The four colours used in the hibualamo house have their meaning. The red colour reflects the tenacity of struggle, the yellow colour represents intelligence and wealth, the black colour represents solidarity, and the white colour represents purity.

4. *Rumah Wetar* - The tropical island of Wetar belongs to the Indonesian province of Maluku and is the largest island of the Barat Daya Islands (literally Southwest Islands). It lies east of the Lesser Sunda Islands, which include nearby Alor and Timor, but it is politically part of the Maluku Islands.

Figure 205 Source: http://www.myislands.pl/?p=6937&lang=en

5. *Kadato Kie*, Kedaton of Tidore[78]

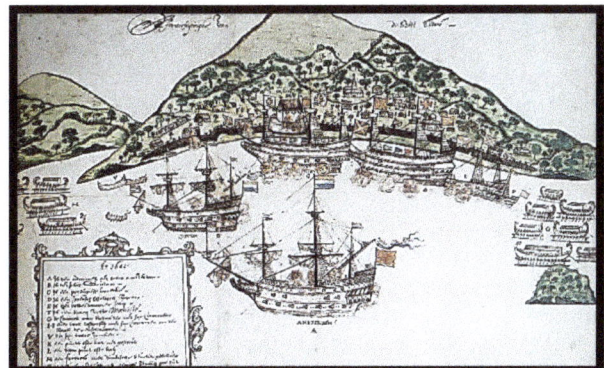

Figure 207 Image of Tidore town in 1601, with Spanish and Dutch ships engaged in a fight. A mosque, a Catholic church and a small fortress can be seen.

Figure 206 Front view of the Sultan Tidore Kedaton

---

[78] Source: https://en.wikipedia.org/wiki/Sultanate_of_Tidore

The Sultanate of Tidore (Indonesian: كسلطانن تيدوري, Kesultanan Tidore, sometimes Kerajaan Tidore) was a sultanate in Southeast Asia, centred on Tidore in the Maluku Islands (presently in North Maluku Province). It was also known as Duko, its ruler carrying the title *Kië ma-kolano* (Ruler of the Mountain). Tidore was a rival of the Sultanate of Ternate for control of the spice trade and had an important historical role in binding the archipelagic civilizations of Indonesia to the Papuan world.

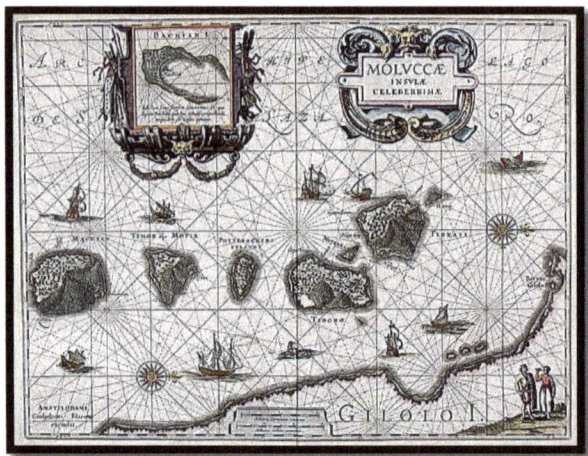 

Figure 208 Left: Early map of northern Maluku made during the Age of Discovery. North is on the right, with Ternate as the rightmost followed by Tidore, Mare, Moti and Makian islands. The bottom is the Gilolo (Jailolo or Halmahera) Island. The inset on the top is Bacan Island. Willem Blaeu, 1630; Right: Entrance to the palace of the sultan of Ternate

6. Sultanate of Ternate[79]

Ternate is a city in the Indonesian province of North Maluku and an island in the Maluku Islands. It was North Maluku's de facto provincial capital before Sofifi on the nearby coast of Halmahera became the capital in 2010.

7. Banda Island - The Banda[80] Islands (Indonesian: Kepulauan Banda) are a volcanic group of ten small volcanic islands in the Banda Sea.

8. Buru Island - The **Wai Apu**[81] people are one of the native peoples of Buru Island in Maluku,

Figure 209 Administrator's house for the nutmeg production 'on Banda

---

[79] Source: https://en.wikipedia.org/wiki/Ternate#/media/File:COLLECTIE_TROPENMUSEUM_Ingang_van_het_paleis_van_de_sultan_van_Ternate_TMnr_60018584.jpg
[80] https://en.wikipedia.org/wiki/Banda_Islands
[81] https://en.wikipedia.org/wiki/Wai_Apu_people

Indonesia, typically inhabiting the northeast of the island in what are now the Namlea and Waplau districts.

Buru people[82] (Indonesian: Suku Buru) are an ethnic group mostly living on the Indonesian island of Buru and some other Maluku Islands. They also call themselves *Gebfuka* or *Gebemliar*, which means "people of the world" or "people of the land". **Kayeli** is an ethnic

Figure 211 Left: Traditional Buru tribe house; Middle: A traditional mosque in Kayeli, circa 1890 to 1940; Right: Lisela House

group living on the southern coast, mainly from the Kaiely Gulf. **Lisela** or **Rana** are an ethnic group mostly living on Buru and some other Maluku Islands.

9. Halmahera Island - The **Tobelo**[83] people are one of the northern Halmahera peoples living in eastern Indonesia, in the northern part of the Maluku Islands and on the eastern side of North Halmahera Regency. They also dominated such small peoples of the interior

Figure 210 A Protestant church in Tobelo, 1924.

of northern Halmahera, such as the Pagu and Tabaru people. The Tobelo people are highly mobile, but their settlements are mainly along the coastline. Ground skeleton-stilted houses (*tathu*) are built from bamboo, and the roofing is made of leaves of sago palms or roof shingles.

Figure 212 Stilt houses on the island of Morotai

---

[82] https://en.wikipedia.org/wiki/Buru_people
[83] https://en.wikipedia.org/wiki/Tobelo_people

10. Morotai Island – Morotai is a rugged, forested island north of Halmahera. The people here build **charnel**[84] houses - structures where the bodies of dead people are kept.

11. **Romang**[85] is an island, part of the Barat Daya Islands in Indonesia. It is included within the Terselatan Islands District (Kecamatan Pulau-Pulau Terselatan) within the Barat Daya Islands Regency of the Maluku Province. Alternate names in use are Roma, Romonu and Fataluku.

Figure 213 Indigenous huts on Romang Island

12. Seram Island - Seram[86] (formerly spelt Ceram; also Seran or Serang) is the largest and main island of Maluku province of Indonesia, despite Ambon Island's historical importance. It is located just north of the smaller Ambon Island and other adjacent islands, such as Saparua, Haruku, Nusa Laut and the Banda Islands.

Figure 214 Rade and village of Warrou (Ceram Island). Picturesque Atlas

Seram Island is remarkable for its high degree of localised bird endemism. There are 117 species of birds on the island, and 14 species or subspecies are endemic. The mammals found on Seram include Asian species (murid rodents) and Australasian marsupials. The mountain area of Seram supports the greatest number of endemic mammals of any island in the region. It harbours 38 mammal species and includes nine endemic or near-endemic species, several of which are limited to montane habitats.

---

[84] https://commons.wikimedia.org/wiki/Category:Morotai
[85] https://en.wikipedia.org/wiki/Romang_(island)
[86] https://en.wikipedia.org/wiki/Seram_Island

Chapter 13

Vernacular architecture of East Nusa Tenggara

Figure 215 Islands of Nusa Tenggara Timur (NTT) East Nusa Tenggara
Source:

We now explore the architecture of these Islands in Nusa Tenggara Timur (NTT). The name Nusa Tenggara is Indonesian for "southeast islands."

East Nusa Tenggara[87] (Indonesian: Nusa Tenggara Timur – NTT) is the southernmost province of Indonesia. It comprises the eastern portion of the Lesser Sunda Islands, facing the Indian Ocean in the south and the Flores Sea in the north. It consists of more than 500 islands, with the largest ones being Sumba, Flores, and the western part of Timor; the latter shares a land border with the separate nation of East Timor. The province is subdivided into twenty-one regencies and the regency-level city of Kupang, the capital and largest city. Source[88]

Flores Island

Figure 216 Vernacular houses of East Nusa Tenggara

---

[87] https://en.wikipedia.org/wiki/East_Nusa_Tenggara#/media/File:Nusa_Tenggara_Timur.png
[88] https://en.wikipedia.org/wiki/East_Nusa_Tenggara

One of the larger major Islands, each with a few ethnic tribes with distinctly different architecture in the traditional villages.

Figure 217 Chief's house in Ruteng Pu'u Village, Manggarai; Conical hut of Wae Rebo village and Gurusina Village;

Figure 218 Left: Wolowaru Village; Right: Wologai Village, Flores Is.

Etymology From Portuguese flores, plural of flor ("flower").

Figure 196 shows **Manggaraian** houses – the middle and left photos are the more ancient roundhouses in Wae Rebo, Todo, and Gurusina Village. Figure 197 is the more modern roundhouse in many villages today (usually as a chief's house/clan house), e.g., in Ruteng Pu'u and villages around Cancar. The Ruteng type is a building on posts. No examples exist anymore where this building type is constructed with walls, and if so, it looks like the posts lean outwards with the walls slanting outwards. Wolowaru and Wologai types have roofs tapered to ridges instead of pointed conical ones. ***Ende***–Wolotopo, a village close to Ende, has two multi-family houses unique to the area. Due to the size of the houses and the location within the village, it was impossible to take photos from the front. The house with the dark roof was pulled down in 2011/2012 and is currently being reconstructed. The house with the lighter roof was a very special type of construction that is not typical for Flores (Figure 199).

Figure 220 The village near Ende town, Flores Island

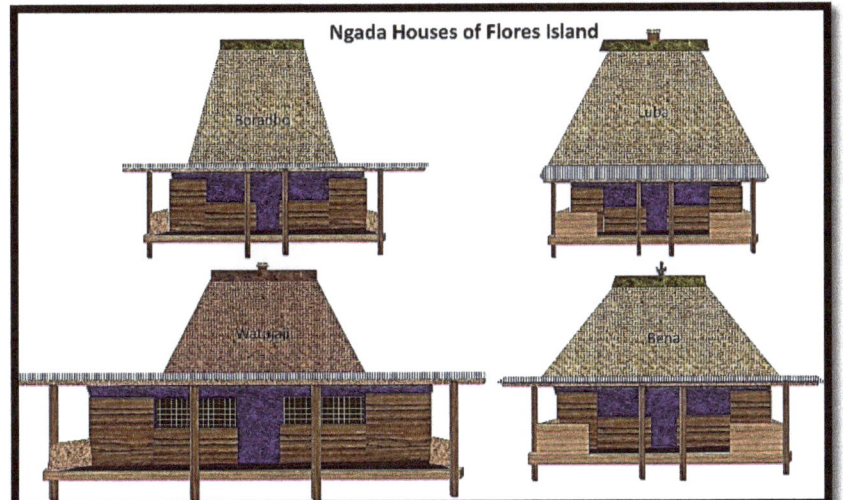

Figure 219 Top left: in Boradho; Top right: in Luba; Bottom left: contemporary version in Watujajii; Bottom right: in Bena

Ngada houses have several different roof shapes, with a single-pitch bamboo roof on the veranda. Some examples from different regions are in Figure 198 Top left: in Boradho; Top right: in Luba; Bottom left: contemporary version in Watujajii; Bottom right: in Bena).

Lopo is a half-round house where Boti people gather, for example, to listen to the directions given by the King of **Boti** (Usif Boti) or just for the usual gathering on the ninth day of the Boti calendar. In terms of shape, *lopo* has similarities to *ume kbubu* or round-shape houses often used by the South-Central Timor community as kitchens, food warehouses, and heaters.

Figure 221 Lopo in Boti Village - South Central

Figure 222 Typical Boti village Hut

Boti Village - Boti, deep in the mountains of Kei country, is regarded as one of the most authentic traditional villages in West Timor. They are direct descendants of the Dawan (*Atoni Metu*),

Timor's first indigenous tribe. *Boti* Village is about 30km east of Soe, Southwest of West Timor. There is also a *Boti* town on Rote Island, South of Timor Island.

From **Boti**[89] village, we divert our attention to **Bena** Village[90], one the most beautiful traditional villages on Flores Island, Indonesia. Currently, Bena Village consists of approximately 45 houses surrounding each other with nine tribes that inhabit these houses, namely the *Dizi* tribe, *Dizi Azi* tribe, *Wahto* tribe, *Deru Lalulewa* tribe, *Deru Solamae* tribe, *Ngada* tribe, *Khopa* tribe and the *Ago* tribe. The difference between one tribe and another is that there are nine levels, and each tribe is in one level of height. The arrangement of the houses in Bena looks unique because their semi-circular shape forms the letter U, and each house also has a roof decoration that is different from one another based on the lineage of the ruling and living in the house.

Figure 223 Traditional hut in the Flores village of Bena

In the middle of the village, there is usually a building that the local people call Bena, *Nga'du* and *Bhaga*. Both are symbols of the village ancestors who are in the yard, *kisanatapat*, where traditional ceremonies are held to communicate with their ancestors. *Nga'du* means a symbol of male ancestors, and its shape resembles an umbrella with a single pole building and a roof of palm fibre so that it looks like a shade hut. The Ngadhu pole is usually made of a special hardwood because it also functions as a gallows for sacrificial animals during traditional celebrations. In comparison, *Bhaga* means a symbol of female ancestors whose shape resembles the shape of a miniature house.

---

[89] W B N Dosinaeng et al. Ethnomathematics in Boti tribe culture and its integration *Journal of Physics: Conference Series 1657 (2020) 012021 IOP Publishing doi:10.1088/1742-6596/1657/1/012021*
[90] Source: https://indonesiakaya.com/pustaka-indonesia/desa-bena-warisan-budaya-zaman-batu-di-bajawa-flores/

Figure 226 Lio chiefs house in Pemo, Lembata Island; Lio chiefs house in Jopu -side in Jopuview and front view; Photo: Irene Doubrawa

The houses of the Lio. Most of the houses either feature a higher/more extreme roof with a narrower ridge (Figure 204 left) or a roof that is not all that high (Figure 204 middle top and right) – this depends on the status of the house (as far as I could find out). The houses in Pema and Jopu have high roofs, and the top of the roof gently slopes outwards. The lines of the roof are elegant curves – this is possible through a special roof construction that allows the use of thin bent rafters. In Figure 204, middle bottom, the buildings of the Lio are constructed on posts.

Figure 225 East-Flores clan-houses around Leworahang.

Figure 225 East-Flores Kawaliwu – a "traditional" house with two vertical elements on top of the roof

East-Flores clan houses[91] around **Leworahang**. This image shows the main features of the East Flores dwelling house, and the bottom image shows the outline of the clan house. One distinct feature of the clan houses is the large piece of wood that forms the ridge and is shaped like a crocodile, as well as the three vertical bamboo poles on the roof ridge.

---

[91] Source: https://floressustainabletravel.blog/architecture/

Komodo National Park

The National Park is a group of small Islands located between Bima and Flores. Pulau Misa is one of the islands with a significant Bajao influence. The Bajao have their roots in Sulawesi and spread from Southeast Mindanao (see Figure 206) to North Borneo to NTT. Below are some of the peculiar Bugis-gable finials of Sulawesi and Flores. There are ferries from Sape Port in Badjo, East Bima, to Komodo NP and Badjo, West Flores. Such is the influence of *Bajao Laut* (sea gipsies) ethnicity in this region. The sea gipsies are true masters of the sea, having roamed the waters of the Coral Triangle between Malaysia, the Philippines and Indonesia for many generations.

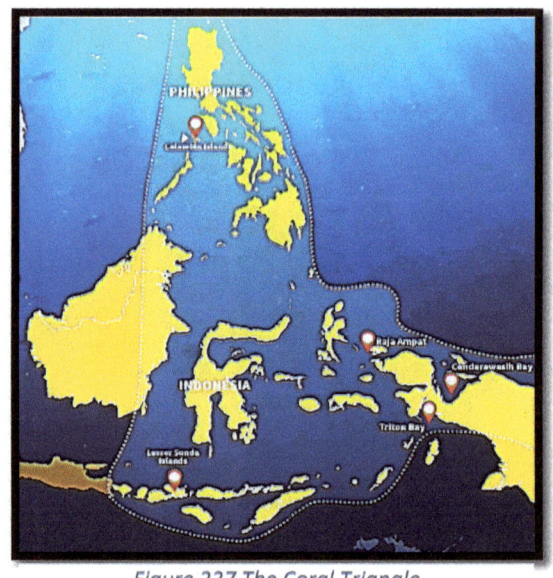

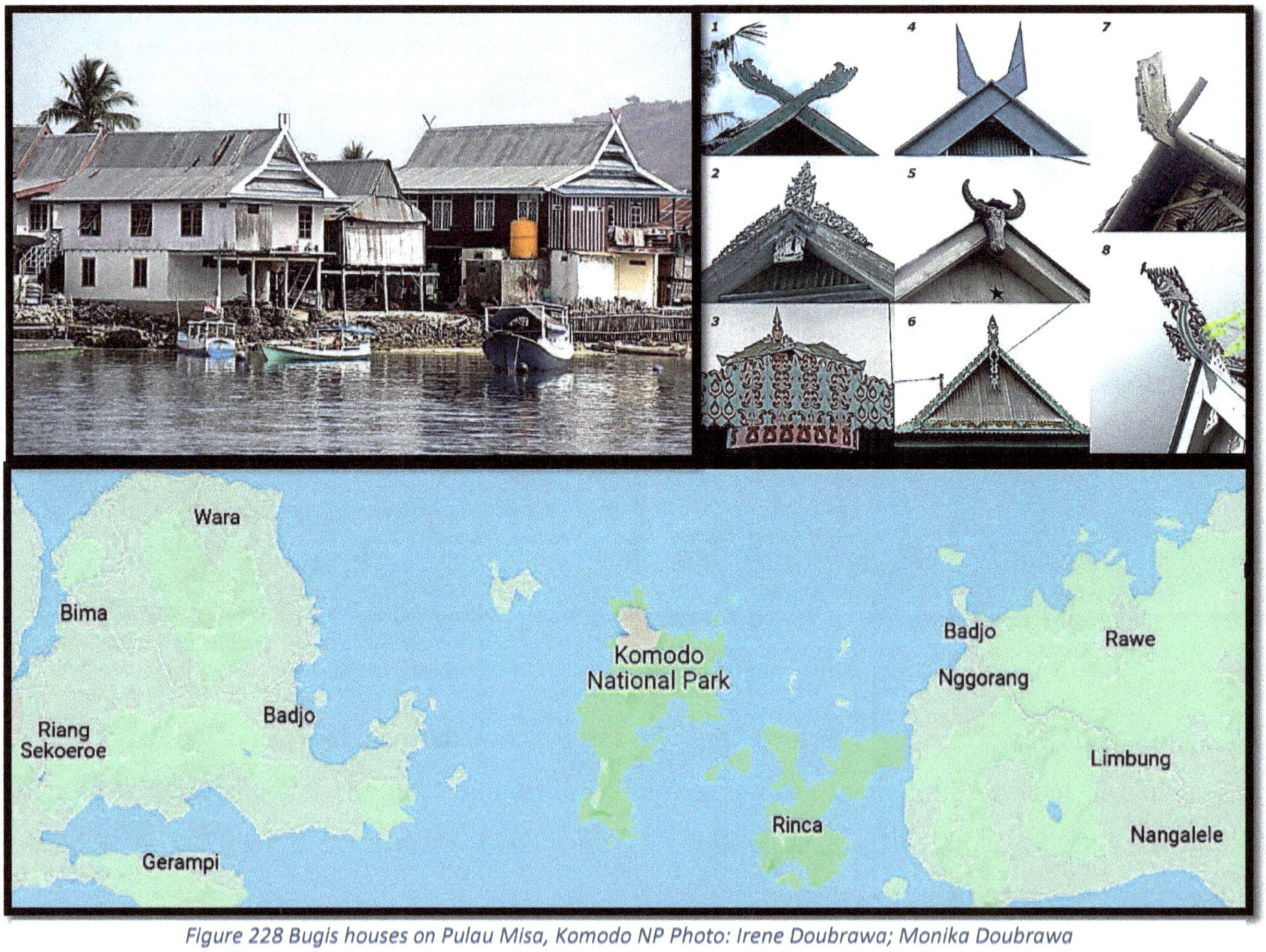

Figure 227 The Coral Triangle

Figure 228 Bugis houses on Pulau Misa, Komodo NP Photo: Irene Doubrawa; Monika Doubrawa

The Sama Bajao Factor

The Sama-Bajau include several Austronesian ethnic groups of Maritime Southeast Asia. The name collectively refers to related people who usually call themselves the Sama or Samah (formally A'a Sama, "Sama people") or are known by the exonym Bajau (/ˈbɑːdʒaʊ, ˈbæ-/, also spelt Badjao, Bajaw, Badjau, Badjaw, Bajo or Bayao). Source[92]:

Figure 229 Bajao Architecture, Traditional ornaments, and boat designs in the Coral Triangle of Malenesian.

---

[92] https://en.wikipedia.org/wiki/Sama-Bajau

Chapter 14

Vernacular Architecture of Sumba and NTT group of Islands

Etymology: According to tradition, the name Sumba derives from Humba. Rambu, or Misses Humba, was the wife of Umbu or Mister Walu Mandoku, one of the chiefs among the first tribes who settled in Sumba. He wanted to perpetuate the name of his beloved wife by naming the island so.

Figure 230 Uma Kelada of Sumba, is an exponential exaggeration of the height of the roof peak as compared to the Javanese Joglo

Uma Kelada is usually found in Sumba Island, East Nusa Tenggara, particularly in Ratenggaro Village. As a part of Sumbanese cultural heritage, the traditional house is known for its towering straw roof, whose height depends on the social status of its inhabitants. Moreover, the foundation of the *Uma Kelada* house is mainly composed of bamboo. No windows are made in this house; instead, the light and air go through the small gaps between the bamboo walls.

In the nearby Sumbawa Island, the towering roof even extends to the extreme of the 'hot air rises' principle. Presumably, the living quarters can be the coolest on the hottest day of the year. Sumbawa, although adjacent, is not in Nusa Tenggara Timur but in Nusa Tenggara Barat (NTB), i.e., West Nusa Tenggara), added here to provide the flow of architectural influence. The other end, where the towering

Figure 231 Typical tapered Welsh hat….; Joseon Period Korean Noble 'Gat' Hat; Quaker Amish Hat or Pilgrim Top Hat

roof evolved from, is the lower peak of the Javanese Joglo (see Figure 128). These towering roofs have some resemblance to old Top Hats[93].

Timor, formerly known as Timor Barat (West Timor) before Timor Leste was declared its own country, was known as Timor Timur (East Timor). Timor & Semau – Kupang, the NTT capital, is the Major port of Timor, and Semau is an Island just west of Kupang. Timor is the name of the Island, while timur means east in Indonesian, although the pronunciation is usually the same.

The Atoni (also known as the Atoin Meto, Atoin Pah Meto or Dawan) are an ethnic group in Timor, in Indonesian West Timor and the East Timorese enclave of Oecussi-Ambeno. Their language is Uab Meto[94].

Figure 232 Forms of the Atoni tribe's traditional houses: Sonaf, Ume Kbubu, Lopo, Ume Kbat//Kanaf

The houses usually form a circular cluster or follow the road after introducing a road.

Ethnonyms[95]: Atoin Pah Meto, Atoin Meto, Timorese; Orang Timor Asli (in Indonesian)

Architectural Form[96]

Timorese people know 'ume kbubu' with different dialects in each region referring to the same house. In the Dawan[97] language, the Amanuban dialect, 'ume' means house and 'kbubu' means round. So *'ume kbubu'* means roundhouse. *Ume kbubu*' is a barn where the

Figure 233 "Ume Kbubu" Traditional East Timorese House that is Environmentally Friendly and Earthquake Resistant

---

[93] https://upload.wikimedia.org/wikipedia/commons/5/55/Typical_tapered_Welsh_hat.jpg
https://upload.wikimedia.org/wikipedia/commons/b/b9/Black_gat3.jpg
https://en.wikipedia.org/wiki/Amish_way_of_life#Clothing
[94] Source: https://en.wikipedia.org/wiki/Atoni
[95] https://www.encyclopedia.com/humanities/encyclopedias-almanacs-transcripts-and-maps/atoni
[96] Sitindjak, H. I. et al., 2020 The Iconography of Sonaf Nis Non-Traditional House in East Nusa Tenggara, Indonesia. *SHS Web of Conferences 76, 01046* https://doi.org/10.1051/shsconf/20207601046
[97] https://www.kompasiana.com/neno1069/5dbd69a9d541df247b61b652/ume-kbubu-rsia-suku-dawan-timor-tinggal-kenangan

people of Timor (atoin meto) store their crops, which are round with four pillars, thatched roofs, and Earthquake Resistant.[98]"

Letmafo Society, North Central Timor District[99]

Maubes traditional house, or *Sonaf Maubes*, is the physical evidence of the existence, identity, and characteristics of Letmafo society. Maubes traditional house is designed as a physical construction inhabitation and has a socio-cultural meaning that covers any life dimensions.

The house stores various historical relics and myth stories. The historical relics and myth stories believe by the villages could give confidence and belief to the owner about something.

Figure 234 The traditional house of Sonaf Maubes

Group of Solor Islands

There are many more small Islands and not significantly called out, possibly due to fewer inhabitants or not being inhabitable. However, Meliponines may still be present and yet undiscovered as the vegetation is there for foraging without human intervention.

East Nusa Tenggara, Indonesian Nusa Tenggara Timur, Propinsi (or provinsi; province) of Indonesia comprising islands in the eastern portion of the Lesser Sunda Islands group: Sumba, Flores, Komodo,

Figure 235 Map of The Solor Island group and the Alor Island Group

---

[98] :https://www.kompasiana.com/sayyidati25722/5bc02081677ffb575477e8f3/ume-kbubu-rumah-traditional-community-timor-yang-environmentally-friendly-dan-earthquake-resistant

[99] Kabosu M.Y. et al. 2018 Maubes Traditional House: The Cultural Legacy of Letmafo Society, North Central Timor District, East Nusa Tenggara. International Journal of Multicultural and Multireligious Understanding Volume 5, Issue 4 August, 241-248 http://ijmmu.com

Rinca, the Solor Islands (Solor, Adonora, and Lomblen), the Alor Islands (Alor and Pantar), Sawu, Roti, Semau, and the western half of Timor.

History of the Solor Islands

It was written in a journal[100] the Portuguese came to Solor Island around 1561 AD. Then, they built Fort Lohayong in 1566 AD. At that time, the people of Solor and its surroundings asked the Sultan of Menanga to lead the resistance against the Portuguese. The Sultan of Menanga's resistance to the Alliance against the Portuguese was supported by the Vereenigde Oostindische Compagnie (VOC), a Dutch trading company.

The main reason why the Portuguese controlled the southern coast of the island of Flores was that controlling southern coast to the island of Flores, the Portuguese could easily sail to the island of Timor, separated only by the Savu Sea, which had white sandalwood, which was very expensive in the European market at that time.

Figure 236 Group Portrait with missionary pastor J. van der Loo in front of the Roman Catholic church in Konga (circa 1915)

The historical Portuguese presence on the island of Timor cannot be separated from the Portuguese presence on the island of Flores and its surroundings; even the first Portuguese colonial centre in Flores, Solor, and Timor was on the island of Solor, in East Nusa Tenggara province, Indonesia before finally Portuguese being moved to Lifau and in 1769 it was moved again to Dili which later became the capital of the state of Timor Leste.

History of the kingdoms of Adonara island

There are three kingdoms on the island of Adonara:

- **Adonara**
- **Terong**
- **Lamahala.**

Together with the 2 kingdoms on the island of Solor (Lamakera and Lohayong), these three kingdoms are called Watan Lima. The descendants are supposed to be Bugis.

Local history on Adonara[101] is documented from the 16th century when Portuguese traders and missionaries established a post on the nearby island of Solor. By then, Adonara and the surrounding

---

[100] Situs Menanga Solor Flores Timur, Jejak Islam di NTT written by Muhammad Murtadlo and published in the Journal of Religious Literature Ministry of Religious Affairs in 2017

[101] Source: https://id.wikipedia.org/wiki/Kerajaan_Adonara
– Video history of the kingdoms of Nusa Tenggara Timur: https://www.youtube.com/watch?v=e8x3xQIr6Xw

islands were ritually divided between a population of coastal dwellers known as Paji and a population mainly settling in the mountainous inland called Demon.

The Paji areas on Adonara contained three principalities: Adonara proper (centred on the island's north coast) and Terong and Lamahala (on the south coast). The Paji were susceptible to Islam, while the Demon tended to fall under Portuguese influence. Together with two principalities on Solor (Lohayong and Lamakera), they constituted a league called Watan Lema ("the five shores").

The Watan Lema allied with the Dutch East India Company (VOC) in 1613, confirmed in 1646. The Adonara principalities had frequent feuds with the Portuguese in Larantuka on Flores and were not always obedient to the Dutch authorities. In the nineteenth century, the ruler of Adonara (proper) in the north strengthened his position in the Solor Archipelago; by then, he was also the overlord of parts of eastern Flores and Lembata. The Demon areas stood under the suzerainty of the principality of Larantuka, under Portuguese rule until 1859, when it was ceded to the Netherlands. The Indonesian government abolished the principalities of Larantuka and Adonara (proper) in 1962.

Etymology: Adonara[102] combines two words in the Lamaholot language (including Adonara): Ado and nara. Ado is the name of the first male to inhabit the island of Adonara, namely Kelake Ado Pehan, while nara means village, nation, or relatives. Adonara means the village of Ado, the tribe of Ado, descendants and relatives of Ado.

Figure 237 Traditional house of Nagekeo; Open-air Museum with a traditional house in Ende. Front and side view

Nagekeo house, a "typical traditional" that has the top of the roof narrowed down towards a very short roof ridge, which is then covered with many layers of alang-alang grass – this accounts for the impression that the top of the roof is slanting outwards, just like an upside-down trapezoid (Sketch 3). Also, the houses are built on stilts (three along the front side of the house). The two vertical elements on top of the roof show that the house is of high status (has a special function within the village/clan) – "normal" houses don't feature these two vertical poles on top of the roof.

---

[102] https://en.wikipedia.org/wiki/Adonara

Solor & Adonara are easterly adjacent to Flores, and both have quite similar cultures and architecture, with Lomblen being more significant among tourists. The Solor house (Figure 218) has similarities to the gable roof style of a Roman Catholic church in Konga (Figure 215), a district in Flores adjacent to Solor Island.

Figure 239 Solor Island House

Figure 239 Lithography of a church in Lomblen
Source:

Lomblen Island is the main island that forms Lembata Regency, East Nusa Tenggara Province, Indonesia. Adonara is located east of the larger island of Flores in the Solor Archipelago. Lomblen, sometimes called Kawela or Lembata, is an island between Adonara Island and between the Nusa Tenggara Islands and Pantar Island.

Lembata Island

Figure 241 Traditional house of Ile Ape Timur, Lembata:

Figure 240 Ile Ape Timur (East Ile Ape), Lembata Island

NuhaNera (NN) Homestay, Lamatokan, Lembata, East Ile Ape, Lewoleba[103], Lembata Island (Figure 220) is in the North and about an hour from Lewoleba Airport. Lamatokan is a rock outcrop in a bay protected from the open Banda Sea. The houses here have Gambrel-type roofs.

Savu & Raijua Savu or Sawu or Sabu is adjacent to Raija, sometimes referred to as **Sabu Raijua**.

Figure 242 Gambrel-type roofs of Homestays in Lewoleba, Lembata Island

Sabunese culture: Sabunese has their language and used to follow an ancestral indigenous religion called jingi tiu. This religion gives a large ritual importance to megalithic stones like the one found in the village of Namata. Sabunese have their traditional architecture and are probably among Indonesia's finest ikat weavers.

Figure 243 Traditional Huts in Top Left: Rote Island; Top Middle: Sika Island; Top Right: Alor Island; Bottom Left: Raijua Island; Bottom Middle and Right: Savu Island

---

[103] https://www.tripadvisor.com/LocationPhotoDirectLink-g2301830-d14057919-i317934121-Lembata_Exotic_Tour-Lembata_Island_East_Nusa_Tenggara.html

Pantar & Alor Island These islands are the easternmost in NTT, with Alor being the more developed and significant in that area. Below are some traditional houses of the Abui Tribe in Takpala Village[104], Alor Island, Indonesia.

Pantar Island

The three main domains on Pantar[105] (Van Galen 1946 and Hägerdal 2010, 28-30) are Pandai, northern, Baranusa, western and Blagar, eastern Pantar.

The landschaps[106] of Blagar and Pandai were merged to form Pantar Matahari Naik in 1918. This merger was ruled by the former Raja of Pandai, Koliamang Wono, until 1926. Below are some traditional houses among the first Papuan people who settled on Pantar. The Papuan languages of Alor and Pantar form a discrete, relatively isolated subgroup of the Timor-Alor-Pantar language family (Schapper et al. 2012)

Figure 246 Rote Island - Beach Hut

Figure 244 Traditional houses of the Abui Tribe in Takpala Village, Alor Island, Indonesia.

Figure 245 Lebang, Central Pantar; Muriabang, Southern Pantar photos by Ernst Vatter in 1929; Right: Port of Baranusa, Western Pantar (Photographed by Professor Robert Barnes in 1996, Shared Shelf Commons)

Rotenese people are one of the native inhabitants of Rote Island, while part of them reside in Timor. Rote Island, an isle south of Semau, is called Roti (although roti means bread in Bahasa Indonesia).

---

[104] Source: https://www.universitybridgeproject.org/blog/tradition-and-culture-in-the-village-of-takpala-alor-island-east-nusa-tenggara-province-indonesia
[105] Source: http://www.asiantextilestudies.com/pantarisland.html
[106] landschap n (plural landschappen, diminutive landschapje n) Dutch (archaic) = region.

The Rotenese people also settled in islands surrounding Rote Island, such as Ndao Island, Nuse Island, Pamana Island, Doo Island, Heliana Island, Landu Island, Manuk Island, and other smaller islands.

The kinship system of these people is nucleus family kinship or broad family with patriarchist in character and maintains an exogamy clan marriage custom.[2] The formation of large families consists of the smaller clans called *nggi leo*; these smaller clans make up the larger clans called leo.

Traditional Rotenese clothing is a *kain* (a cloth of up to 2.5 meters long, wrapped around the waist, reaching to the knees or ankles), as well as jackets and shirts, with a specific style of a straw hat called *ti'i langga*.

A conical design (Figure 226) in the jungles of Rote Island South of Timor. This conical design is inspired by a bed and breakfast accommodation in the jungles of Rote Island. Although Rote Island is better known as a surfer's paradise with excellent beach fronts, the jungle retreat is also lovely.

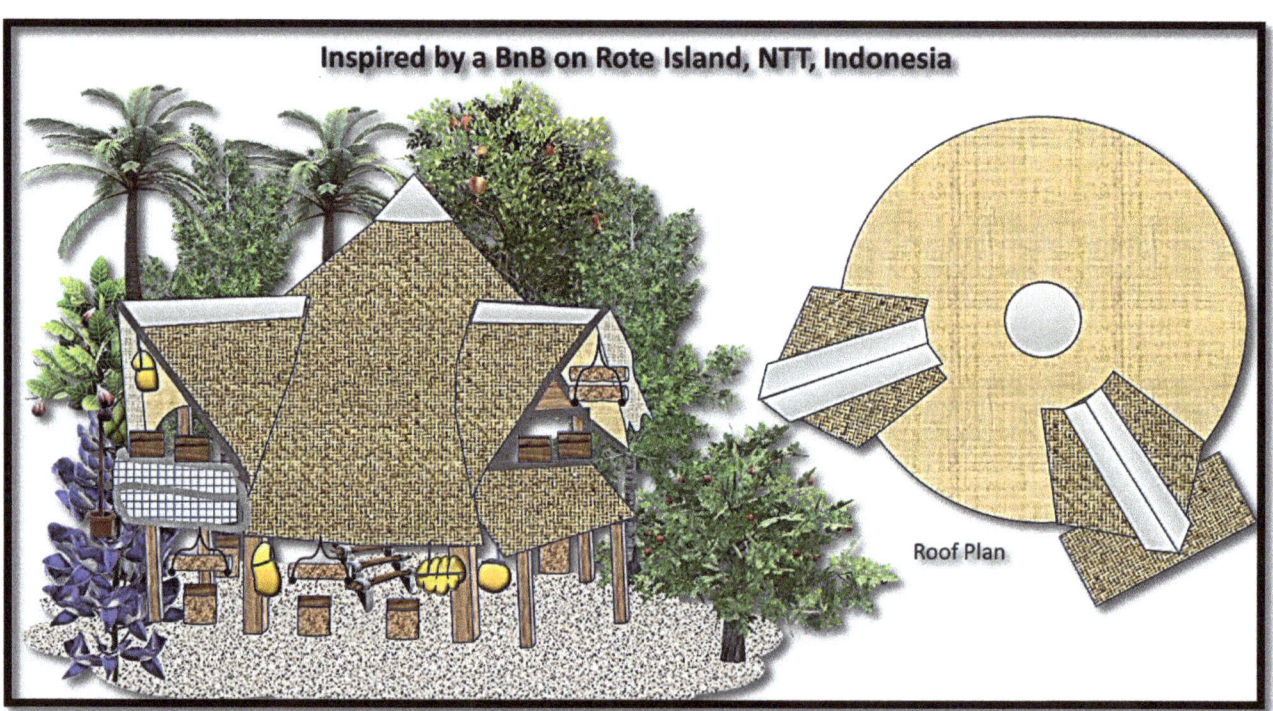

Figure 247 Inspired by a BnB on Rote Island, NTT, Indonesia

Figure 248 Left: Sika Island Beach Huts; Right: Sika Island - Dugong watching post

Sika is a tiny island east of the Alor Archipelago and North of Timor Island. Dugong watching post has a unique vernacular project conical hut.

The Sikka (also Sikkanese, Sika) people are an Indonesian ethnic group native to the region of east central Flores between the Bloh and Napung Rivers. In Maumere, the region's centre, Sikka people occupy a separate block.

Timorese Sika - They retained their original Malay language but later switched to Creole Portuguese. Today, they have been absorbed into the same population and no longer form their distinct group. The primary religion practised by the Sika people is Roman Catholicism. Sika people residing in the interior still retain their traditional ancestral worship and Agrarianism cult.

Mountain villages are small and have a circular layout. They are located on the steep slopes of the mountains, which serve as protection against attacks. In the middle of the settlement is an area with a temple and sacred megalithic shrines. Coastal settlements have a linear plan located along a road or river. The dwelling frame and pillar structure, pile, in the mountains, is designed for large families, while in the coastal areas, for a small family.

Addendum Gallery

Figure 249 Barn Type Beehive Shelter

Ideas Gallery for Vernacular Bee Housing, Racks, Shacks & Sheds designs.

Figure 250 Model replica roof of vernacular Sumatran houses. Top: Minangkabau; Bottom: Batak Toba built by Heri Damora.

Figure 252 Balinese architecture inspired Bee Rack

Figure 251 Some novelty ideas found in Lampung by Heri Damora. Top: Model Truck Bee Box; Bottom left Model ship beehive; Bottom right a Bee tower model of a condominium

Figure 254 replica of a Toraja house in Sulawesi built by Heri Damora.

Figure 253 Bee racks Inspired by the Kalimantan tribal communal longhouse. Rack sketch by Abu Hazqeiq.

Figure 256 Vernacular architecture in Bandung, Java. A replica with the monitor roof for a bee box hive was built by Nzank Bandung. 3rd angle projection with an exploded view.

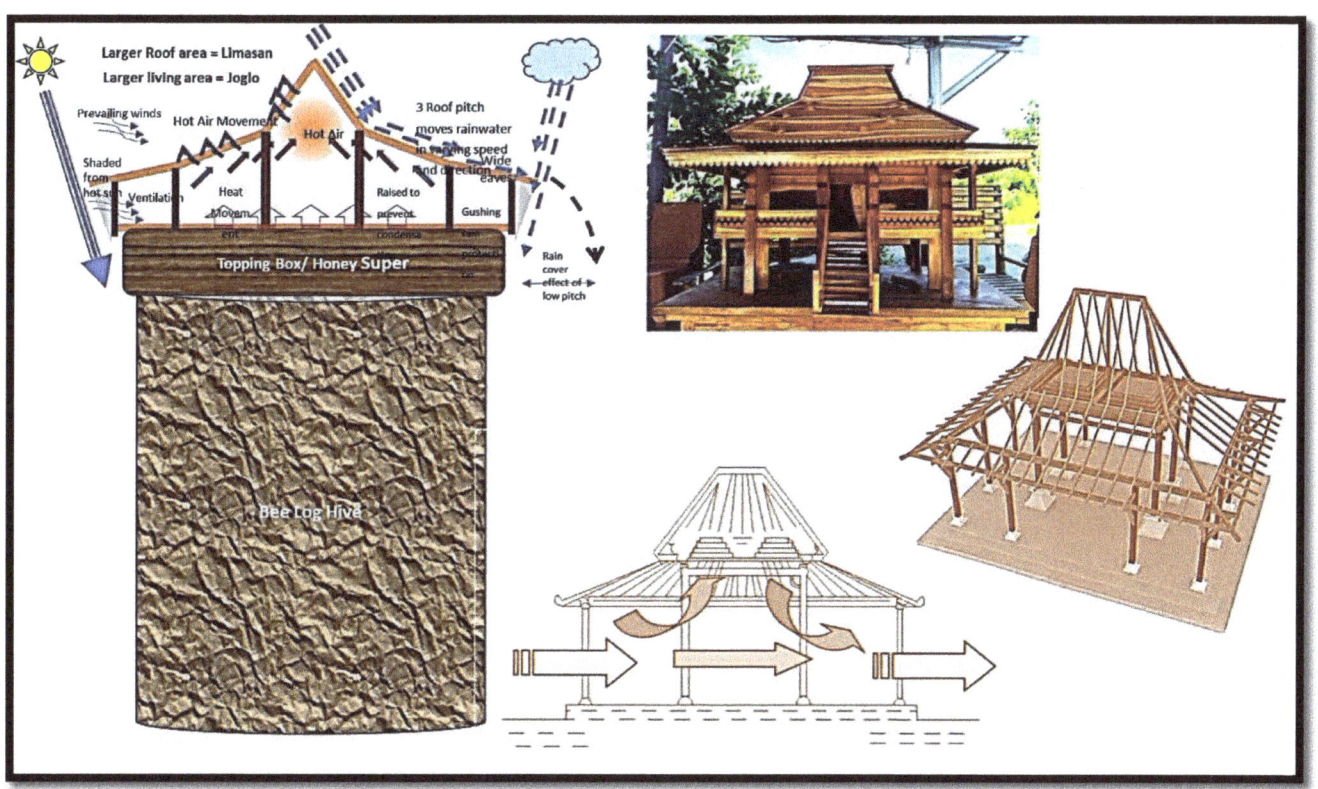

Figure 255 The Javanese traditional Joglo house replica built by Heri Damora.

List of Figures

Figure 1 Vernacular Fusion – Batak Karo roof on a West Kalimantan tall house. 3
Figure 2 Beyond Reality – Power of Symmetry. 10
Figure 3 Rolled-up Bamboo blinds with a SB nest within. 10
Figure 4 Assorted Miniature models and bottom right: En. Baser Ali at work 12
Figure 5 More models and the bottom is the Melaka Palace replica 13
Figure 6 Conservation in Meliponiculture 15
Figure 7 Examples of Bee housing with vernacular roof designs. Image from the back cover of Beescape Book 2014 16
Figure 8 The Malay World in the Indo-Malaya Ecozone 17
Figure 9 Passive ventilation and airflow in a Malay traditional house 19
Figure 10 Typical parts of a Malay vernacular roof 20
Figure 11 Various rake boards for a Kelantan House 21
Figure 12 Rumah Lipat Kajang, a type of traditional Riau house with tiled stairs in the Taman Mini Indonesia theme park. Source: https://www.wikiwand.com/en/Malay_house 22
Figure 13 Cross-section of a Malay House in the 40s to 60s. 23
Figure 14 Typical Terengganu House 24
Figure 15 Inspired by the Rumah Kutai or Rumah Warisan of Perak, Malaysia. 25
Figure 16 In Perak, the long-roof house is known as the Kutai House and Rumah Potong Pattani. 25
Figure 17 Rumah Perak 25
Figure 18 Tetragonula basimaculata (Bingham 1903) 26
Figure 19 Tetragonula melina (Gribodo, 1893) 26
Figure 20 In Selangor and the Federal Territory, the long-roof houses were influenced by the architecture of the long-roof houses in Melaka and Negeri Sembilan. 27
Figure 21 Tetragonula minor (Sakagami, 1978) 27
Figure 22 Rumah Melaka 28
Figure 23 In Melaka, the long roof house, also known as Rumah Serambi Melaka, is divided into a house of twelve pillars and a house of sixteen pillars. 28
Figure 24 Tetragonilla atripes (Smith, 1857) 28
Figure 25 Rumah Serambi Gajah Menyusu – Most of them are found in Pulau Pinang. 29
Figure 26 Tetragonula fuscobalteata (Cameron, 1908) 30
Figure 27 Tetragonula zucchii (Sakagami, 1978) 30
Figure 28 Tetragonula testaceitarsis (Cameron, 1901) 30
Figure 29 In Kedah and Perlis, Malay houses with long trunks are known as Rumah Serambi Kedah @ sometimes called longhouses only. 31
Figure 31 The long roof that is Minangkabau 32
Figure 31 Rumah Warisan Negri Sembilan 32
Figure 33 Rumah Pisang Sesikat 32
Figure 33 Rumah Pahang 32
Figure 34 Left Palembang Limas House Source: https://www.wikiwand.com/en/Malay_hous ;Right: Johor Limas House 33
Figure 36 Rumah Limas Johor (Gable and Hip Roof) 33
Figure 36 Johor: Rumah Potong Belanda @ Rumah Limas 33
Figure 37 Houses in Kelantan, there are two types of long-roofed houses which are Rumah Tiang Dua Belas and Rumah Bujang 34
Figure 38 Rumah Perlis 34
Figure 39 Terengganu Malay House 35

Figure 40 In Terengganu, there are two types of long-roof houses, namely Rumah Tiang Dua Belas and Rumah Bujang. 35
Figure 42 A traditional Malay Sarawak house in Malaysia. Source: https://www.wikiwand.com/en/Malay_house 36
Figure 42 Traditional Malay Sarawak house 36
Figure 43 Borneotrigona hobbyi (Schwarz, 1937) 37
Figure 44 Tetragonula fuscobalteata (Cameron, 1908) 37
Figure 46 A large Malay Hall originated from the Fold Kajang style and continued with the Limas roof style, Pavilion Riau, and Taman Mini Indonesia theme park. This style of structure is often used in the architecture of the palaces of Malay kings, and royal buildings. Source: https://www.wikiwand.com/en/Malay_house .. 38
Figure 46 Rumah Lancang or Rumah Lontik with a curved roof and a bot-like structure. A traditional Riau Malay house, this is the theme park of the Riau Pavilion Taman Mini Indonesia. Source: https://www.wikiwand.com/en/Malay_house 38
Figure 48 Traditional Bruneian Malay houses on stilts in Kampong Ayer, the traditional riverine settlement in Brunei. Source: https://www.wikiwand.com/en/Malay_house 39
Figure 48 Malay house in Sungai-liat, Bangka Island. Source: https://www.wikiwand.com/en/Malay_house 39
Figure 49 Left: Figure 39 Rumah Panggung; Right: Rumah Bujang Berserambi redrawn from Source: https://www.cgtrader.com/3d-models/architectural/other/rumah-bujang-berserambi-selasar 39
Figure 51 Rumah Belah Bubung - Most religions are found in the Riau Archipelago. 39
Figure 51 Rumah Kejang Lako in Rantau Panjang, Jambi 39
Figure 52 Rumah Panjang Kadazan Sabah or Sabah Kadazan Longhouse 40
Figure 53 Inspired by the traditional Lotud tribal house... 40
Figure 54 Lotud house in the Heritage Village of Kota Kinabalu 40
Figure 55 Model of a traditional Sabah House 41
Figure 56 Inspired by the Murut Longhouse of Sabah. 41
Figure 57 Vernacular structures in North Peninsula Malaysia 43
Figure 58 A typical Orang Asli stilt house in Ulu Kinta, Perak. 43
Figure 59 The framework of a traditional native house... 43
Figure 60 Melanau Traditional House 44
Figure 61 Kg TambulianTraditional houses in Sabah... 45
Figure 62 Melaka Literature Museum, Melaka City, Melaka, Malaysia. Source: https://commons.wikimedia.org/wiki/File:Melaka_Literature_Museum.jpg 46
Figure 63 Rumah Adat Beliting: Rumah Rakit / Rumah Gede 46
Figure 64 Rumah Brunei (Brunei House) in the Heritage Park of Sabah Museum..................................... 46
Figure 65 Ornately carved facade of bee box hives in Langkawi Island. 47
Figure 66 Selected Bee Box hives model Malay House with ornate facade – compilation from anonymous beekeepers 48
Figure 67 Traditional Malay wood carving used in box hive designs. Left: Terengganu Malay House; Middle and right: Lampung community hall design. Those on the right, although built-in Lampung, South Sumatra, the facade is of a motif derived from an ancient Javanese royal emblem and used in South Sumatra. 49
Figure 68 Cross-hipped roof applied on Bee rack and tri combo box hives with roof plan. 49
Figure 69 Cross Gable roof models of box hives. From left: Omjol of Aceh, Middle; Anonymous; Right: Zaidi Ibrahim of Terengganu. 49
Figure 70 Interior of his bee gallery with ornate woodcarvings 50
Figure 71 Izhar Johari's house turned bee gallery which he said was inspired by a Terengganu house 50
Figure 72 Painstaking, he patiently completed this bee gallery over two years.. 51
Figure 73 One of Izhar's earlier box hive roofs.. 52
Figure 74 Malay stylized roof for a box hive 52

157

Figure 75 Some of Izhar's hilt and sheath carvings of Malay daggers. 53
Figure 76 Some of the touristy artefacts in the Gallery. 53
Figure 77 Outdoor hive. 54
Figure 78 Traditional Kelantanese structures at Min House Camp in Kubang Kerian, Kelantan. 54
Figure 79 Collection of bee box hives in different styles for different species in the gallery. Bees forage through the louvred windows. 54
Figure 80 Wan Noriah Wan Ramli, owner of Min House Camp in her Bee Gallery. 55
Figure 81 Bee hives are placed facing the glass louvres so that the bees can fly out to forage. 55
Figure 82 *Alhamdulillah, I had the opportunity to stop by to visit my friends at BBH Honey Gallery in Marang, Terengganu. I was impressed with the spirit shown by my friends. We wish them great success in the future. You can get honeybee honey and kelulut homey from him. Images and captions by Mohd Razif Mamat. May 2022.* 56
Figure 83 Vernacular architecture in the Malay Peninsula. 57
Figure 84 Langkawi Island Malay house and a replica roofing on a log hive in the Arts and culture centre backyard in Langkawi. 58
Figure 85 Malay Nusantara traditional architecture redrawn from an anonymous website of public domain 58
Figure 86 West Wing of Bandung Institute of Technology (ITB) image from Wikimedia. 59
Figure 87 Illustration of the Box hive model in Meliponary belonging to Nzank (pronounced Enjang) in Bandung, Java 59
Figure 88 Exploded sequence of the Bandung model house with a Monitor roof 60
Figure 89 Figure 71 Jungle Trek Camp in Chiang Mai, Thailand, with a row of Kampung houses. 60
Figure 90 Distribution of the Malay village house architecture in the Indo-Malaya ecozone. 61
Figure 91 Example of Bees racks inspired by the Traditional tribal ethnic long house 61
Figure 92 Model Bee houses in Langkawi by Aman Londah 62
Figure 93 Some Malay house model beehives for stingless bees made by Luna Ahcmad Cahayani 62
Figure 94 Istana Kampong Glam right before restoration in August 2001 63
Figure 95 Tetragonula geissleri (Cockerell, 1918) 64
Figure 96 Tetragonula valdezi (Cockerell, 1918) 64
Figure 97 Meliponiculture in Tutong by Mitasby Amit in Brunei Darussalam 65
Figure 98 His Meliponary shed and Front view of Meliponine farm at Mitasby's house in Tutong, Brunei Darussalam. 66
Figure 99 a) b) c) Topping Box hive designs d) Peculiar Nest entrances collection 66
Figure 100 a) b) f) Collection of box hive designs d) dark honey c) e) H. itama hive giving dark honey i) Collection of different colours of meliponine honey. 68
Figure 101 Some assorted vernacular architecture in the Indonesian Archipelago, Compilation of ethnic houses from an Indonesian Tourism website. 72
Figure 102 A Toba Batak house and a model Batak roof of a stingless bee box attached to a log hive. 73
Figure 103 The upswept ridge of a centra Sumatra house model and the Acehnese house model of a bee box hive. On the right is a structure of a saddle roof model usually found in Sumatra. 73
Figure 104 Batak Karo House, N. Sumatra 73
Figure 105 Left: Traditional longhouses at a Karo village near Lake Toba, circa 1870; Middle: A Karo people church affiliated with Karo Batak Protestant Church (GBKP). Kabanjahe, Karo Regency, North Sumatra; Right: Musical instruments and other items identified as Karo Batak, photograph (circa 1870) by Kristen Feilberg. <: 74
Figure 106 Traditional House of The Gayo Lues tribe 75
Figure 107 Traditional House of Gayo Serbejadi 77
Figure 108 Traditional House of Batak Gayo Deret 78
Figure 109 Traditional House of The Batak Kluet tribe 79

Figure 110 Traditional House of Angkola Batak tribe 80
Figure 111 Simalungun traditional house 80
Figure 112 The Mandailing traditional house 80
Figure 113 The Bolon house is a symbolic identity of the Batak people 81
Figure 114 The Pakpak/Dairi traditional house 81
Figure 115 A model of a traditional jambur 81
Figure 116 Vernacular architecture in Sumatra, Indonesia 82
Figure 117 Rumoh Aceh image source: Wikimedia 83
Figure 118 In the background, Aceh Museum featuring the traditional Rumoh Aceh is a modern interpretation of the Cakra Donya roof. Source: https://en.wikipedia.org/wiki/Rumoh_Aceh#/media/File:Museum_Aceh.JPG 84
Figure 119 Acehnese traditional house in Piyeung Datu village, Montasik district, Aceh Besar regency. 84
Figure 120 On the left photo is an Aceh Bee Box Hive designed by Omjol of Peureulak, East Aceh. The photos on the right are those of the Rumah Adat, a communal house at the museum in Banda Aceh, North Sumatra. 85
Figure 121 These are adaptations of modern architecture; one was at their museum, and the other is a Belltower. 85
Figure 122 Bagas Godang in Panyabungan, Mandailing Natal 85
Figure 123 AHJ upon arrival at the Minangkabau airport in Padang, West Sumatra 86
Figure 124 Simulated air movement on a skeletal Minangkabau 'Rumah Gadang' roof design on a box hive model 86
Figure 125 Left: A model bee box hive roofing; Middle: Two rangkiang c. 1895 rice granaries; Right: A model of a rangkiang 87
Figure 126 Left: A wall-less balairung in Batipuh.' Right: A balairung in Matur 88
Figure 127 An Uma, the traditional communal house of the Mentawai 88
Figure 128 Replica of Malay Lontiak House of Kampar Majo Tribe and applied on the roof of a bee box hives rack 89
Figure 129 Lampung traditional house named Nuwou Sesat 91
Figure 130 An icon of Lampung's traditional house, Sesat Balai Agung 92
Figure 131 Rumah Nuwou Balak (Rumah Kepala Suku) 92
Figure 132 Rumah Adat Lampung Nuwou Lunik 93
Figure 133 Teluk Betung in the 1930s 93
Figure 134 Kuano Clan Traditional House and a Model of a traditional beehive house 94
Figure 135 A sketch of the Rumah Ulu of the Uluan people of South Sumatra displayed in the Balaputradeva Museum 94
Figure 136 AHJ family photo 1957 95
Figure 137 Left: Actual Betawi kebaya House; Middle: Model house to accommodate bee box; Right: Finished model built by Heri Damora in Lampung 95
Figure 138 Illustration of weather protection of a Stingless bee topping box on a log hive with a Joglo-type roof. 96
Figure 139 Examples of the Javanese Joglo, Limasan and Kampung roofs 96
Figure 140 Roof configurations redrawn from anonymous public domain 97
Figure 141 Various versions of Joglo and Limasan in Central Java 97
Figure 142 Limasan Pavilion (Pendopo in Java) type roofing 98
Figure 143 The Javan Kampong roof is also found in South Sumatra. On the left is a model bee box hive roof built by Heri Damora. 98
Figure 144 Javanese Joglo model by Heri Damora/ Middle photo shows the simulated air circulation and the right photo is the structure with rafter placement. 98

Figure 145 View of Prawita Garden Bee Gallery at the upper level. 99
Figure 146 The bee gallery in Prawita Garden in Banyumas, Java, visited by local tourists. 99
Figure 147 Assembling the Sundanese model with the saddle roof and scissors clamp finial/ 101
Figure 148 Left: Betawi Joglo in Java; Right: West Javan Village hut model 101
Figure 149 Rumah Adat or traditional house 102
Figure 150 Banten Province Traditional House is known as a Rumah Kasepuhan. Right: Beekeepers' box hive roof 103
Figure 151 West Java / Sunda Province Traditional House is also known as Rumah Kesepuhan Right: Beekeepers' box hive roof 103
Figure 153 Overlapping gable roof was adopted on Bee box hives in Lampung by Heri Damora 104
Figure 153 A Beekeeper's home in Banyumas, Java. 104
Figure 154 *Lepidotrigona sp specimen in the custody of Pak Teguh, Prawita Garden, Ajibarang, Banyumas* 104
Figure 155 Model of a traditional Bawean home: 105
Figure 156 Banyunibo located in the center of paddy field southeast of Ratu Boko. 105
Figure 157 Distribution of Meliponiculture of Tetragonula spp. in Java & Bali 106
Figure 158 Stingless Bee Farm in Bali 106
Figure 159 On the left is an ancestral shrine and on the right pic is a garden lantern. 107
Figure 160 From Left Mrs. Mudra, Felicia Tjua Kustiono, Dr. Bajaree Chuttong and Ms. Norita W. Pangestika, Pak Mudra's mother, AHJ, Pak Mudra's father, Pak Mudra, and Daniel Kustiono. 107
Figure 161 Examples of thatch roofing in Balinese styles done by beekeepers. Right: Traditional Balinese walled residential compound belonged to a common man. 108
Figure 162 The bee housing roofs here were done in South Kalimantan (by a non-Balinese), replicating a Balinese Pavilion in Ubud, Bali with a touch of oriental flair. 108
Figure 163 Gambrel roof, more likely Dutch colonial influence as a Barn roof, and in Lombok, the attic is used to store grain. 109
Figure 164 An experimental Terrarium with a T. fuscobalteata hive in a branch hollow. A small miniature replica souvenir that can easily fit a tiny bee eduction. The 25cm x 13cm miniature version Lumbung for the meliponarium (Meliponine Terrarium). 109
Figure 165 A Lumbung replica model bee box hive by Hasan Asri of Makassar. 110
Figure 166 Traditional home of the Sasak tribe in Lombok 110
Figure 167 The oldest mosque dating from 1634 in Bayan. 110
Figure 168 Dried Maja fruits and gourds turn in kelulut hives. 111
Figure 169 Rumah Budaya Majapahit, Trowulan, East Java 111
Figure 170 An old painting of a Majapahit house with an inset of a Majapahit shadow puppet scene. 111
Figure 171 Inspired by the Sasak Traditional house, N. and Central Lombok Is. 111
Figure 172 Celebrations in 1910-1940 miniature display houses comparing the image on the right of a full-scale house as tall as a coconut tree. 112
Figure 173 Depiction of air movement on a skeletal structure and a box hive model Toraja roof 113
Figure 174 This bee house is inspired by the Dulohupa traditional house in Gorontalo, N. Sulawesi. 115
Figure 175 Wallacetrigona Engel and Rasmussen, new genus replacing Geniotrigona incisa (Sakagami & Inoue, 1989) 116
Figure 176 Vernacular architecture Sulawesi, Sumbawa, Maluku & Papua.00 117
Figure 177 Menado Traditional House: Rumah Pewaris 117
Figure 178 Laika House, Kendari, SE Sulawesi 118
Figure 179 Banua Layuk Traditional House of West Sulawesi / Mamuju Regency. 118
Figure 180 Banua Sura Rumah Adar Mamasa 118
Figure 181 A traditional Bugis dwelling in South Sulawesi 119
Figure 182 "Tarempa" en kantoor, Poelau Sangihe 119

Figure 183 Buton, Sulawesi (region) 119
Figure 184 Banjar Bubungan Tinggi House 120
Figure 185 Baloy Traditional House of North Kalimantan Source: 121
Figure 186 Lamin House of East Kalimnatan Dayak 121
Figure 187 Rumah Undur Undur, Ketapang, Kalimantan 121
Figure 189 Rumah Betang, Dayak Ngaju Longhouse, Central Kalimantan 122
Figure 189 Tapered Bee box hive topping a log hive. 122
Figure 190 Rumah Betang, a traditional Ma'anyan house in Muara Bagok, East Barito Regency, Central Kalimantan 122
Figure 191 A traditional Bidayuh 'baruk' roundhouse in Sarawak, Malaysia. It is a place for community gatherings. Source: https://en.wikipedia.org/wiki/Bidayuh 123
Figure 192 Left: A sandung in Central Kalimantan ca. 1915.; 123
Figure 193 Trigona melanocephala Gribodo 1893 124
Figure 194 Vernacular architecture in Borneo, Java, Bali & Lombok Island. 125
Figure 195 Initially, Ozie built blue bee housing with Balinese-inspired architecture, and as his clientele grew from mostly Singapore Chinese, he changed to oriental red. 126
Figure 196 Left: Meliponary in Barabai west of Banjarmasin city, South Kalimantan; Middle: A totem in front of Ozie's house. Right: Ozie and his family are dressed in Banjar Dayak traditional costume in front of his Dayak house design. 126
Figure 197 Traditionally Sumbawa Island 127
Figure 198 Kupang Traditional House: Sao ata mosa lakitana (Musalaki) House 127
Figure 199 Rumah Adat Musalaki NTT 127
Figure 200 Uma Lengge - Rumah Bima / Suku Dompu 128
Figure 201 Left: Uma Mbolo; Middle: Uma Jompa & Right: Uma Panggu 128
Figure 202 Maluku / Ambon Traditional House: Baileo 129
Figure 203 Inspired by the 8-sided Pavilion of the Sasadu people on N. Maluku. 130
Figure 204 Rumah Adat Maluku Hibualamo (photo credit: abdulromli.it.student.pens.ac.id) 130
Figure 205 Source: http://www.myislands.pl/?p=6937&lang=en 131
Figure 206 Front view of the Sultan Tidore Kedaton 131
Figure 207 Image of Tidore town in 1601, with Spanish and Dutch ships engaged in a fight. A mosque, a Catholic church and a small fortress can be seen. 131
Figure 208 Left: Early map of northern Maluku made during the Age of Discovery. North is on the right, with Ternate as the rightmost followed by Tidore, Mare, Moti and Makian islands. The bottom is the Gilolo (Jailolo or Halmahera) Island. The inset on the top is Bacan Island. Willem Blaeu, 1630; Right: Entrance to the palace of the sultan of Ternate 132
Figure 209 Administrator's house for the nutmeg production 'on Banda 132
Figure 210 A Protestant church in Tobelo, 1924. 133
Figure 211 Left: Traditional Buru tribe house; Middle: A traditional mosque in Kayeli, circa 1890 to 1940; Right: Lisela House 133
Figure 212 Stilt houses on the island of Morotai 133
Figure 213 Indigenous huts on Romang Island 134
Figure 214 Rade and village of Warrou (Ceram Island). Picturesque Atlas 134
Figure 215 Islands of Nusa Tenggara Timur (NTT) East Nusa Tenggara Source: 135
Figure 216 Vernacular houses of East Nusa Tenggara 135
Figure 217 Chief's house in Ruteng Pu'u Village, Manggarai; Conical hut of Wae Rebo village and Gurusina Village; 136
Figure 218 Left: Wolowaru Village; Right: Wologai Village, Flores Is. 136

Figure 219 Top left: in Boradho; Top right: in Luba; Bottom left: contemporary version in Watujajii; Bottom right: in Bena 137
Figure 220 The village near Ende town, Flores Island 137
Figure 221 Lopo in Boti Village - South Central Timor 137
Figure 222 Typical Boti village Hut 137
Figure 223 Traditional hut in the Flores village of Bena 138
Figure 225 East-Flores Kawaliwu – a "traditional" house with two vertical elements on top of the roof 139
Figure 225 East-Flores clan-houses around Leworahang. 139
Figure 226 Lio chiefs house in Pemo, Lembata Island; Lio chiefs house in Jopu -side in Jopuview and front view; Photo: Irene Doubrawa 139
Figure 227 The Coral Triangle 140
Figure 228 Bugis houses on Pulau Misa, Komodo NP Photo: Irene Doubrawa; Monika Doubrawa 140
Figure 229 Bajao Architecture, Traditional ornaments, and boat designs in the Coral Triangle of Malenesian. 141
Figure 230 Uma Kelada of Sumba, is an exponential exaggeration of the height of the roof peak as compared to the Javanese Joglo 142
Figure 231 Typical tapered Welsh hat….; Joseon Period Korean Noble 'Gat' Hat; Quaker Amish Hat or Pilgrim Top Hat 142
Figure 232 Forms of the Atoni tribe's traditional houses: Sonaf, Ume Kbubu, Lopo, Ume Kbat//Kanaf 143
Figure 233 "Ume Kbubu" Traditional East Timorese House that is Environmentally Friendly and Earthquake Resistant 143
Figure 234 The traditional house of Sonaf Maubes 144
Figure 235 Map of The Solor Island group and the Alor Island Group 144
Figure 236 Group Portrait with missionary pastor J. van der Loo in front of the Roman Catholic church in Konga (circa 1915) 145
Figure 237 Traditional house of Nagekeo; Open-air Museum with a traditional house in Ende. Front and side view 146
Figure 239 Solor Island House 147
Figure 239 Lithography of a church in Lomblen 147
Figure 240 Ile Ape Timur (East Ile Ape), Lembata Island 148
Figure 241 Traditional house of Ile Ape Timur, Lembata: 148
Figure 242 Gambrel-type roofs of Homestays in Lewoleba, Lembata Island 148
Figure 243 Traditional Huts in Top Left: Rote Island; Top Middle: Sika Island; Top Right: Alor Island; Bottom Left: Raijua Island; Bottom Middle and Right: Savu Island 148
Figure 244 Traditional houses of the Abui Tribe in Takpala Village, Alor Island, Indonesia. 149
Figure 245 Lebang, Central Pantar; Muriabang, Southern Pantar photos by Ernst Vatter in 1929; 149
Figure 246 Rote Island - Beach Hut 149
Figure 247 Inspired by a BnB on Rote Island, NTT, Indonesia 150
Figure 248 Left: Sika Island Beach Huts; Right: Sika Island - Dugong watching post 151
Figure 249 Barn Type Beehive Shelter 152
Figure 250 Model replica roof of vernacular Sumatran houses. Top: Minangkabau; Bottom: Batak Toba built by Heri Damora 152
Figure 251 Some novelty ideas found in Lampung by Heri Damora. Top: Model Truck Bee Box; Bottom left Model ship beehive; Bottom right a Bee tower model of a condominium 153
Figure 252 Balinese architecture inspired Bee Rack 153
Figure 253 Bee racks Inspired by the Kalimantan tribal communal longhouse. Rack sketch by Abu Hazqeiq. 154
Figure 254 replica of a Toraja house in Sulawesi built by Heri Damora. 154

Figure 255 The Javanese traditional Joglo house replica built by Heri Damora. 155

Figure 256 Vernacular architecture in Bandung, Java. A replica with the monitor roof for a bee box hive was built by Nzank Bandung. 3rd angle projection with an exploded view. 155

Index

A

aborigines, - 42 -
Abui, - 148 -
Aceh Besar, - 82 -
Aceh Tamiang, - 74 -, - 75 -
Acehnese, - 82 -, - 83 -
Adonara, - 144 -, - 145 -, - 146 -
A-Frame, - 39 -
Ago, - 137 -
air conditioning, - 15 -
air movement, - 86 -
Alan Lightman, - 9 -
Alor, - 130 -, - 144 -, - 148 -
Aluk To Dolo, - 111 -
Aman Londah, - 61 -
Amanuban, - 142 -
Ambon, - 128 -
Ancestral Veneration, - 111 -
Angkola, - 79 -
animistic belief, - 111 -
Arabica coffee, - 76 -
Architectural design, - 15 -
Aren, - 99 -
Arenga, - 109 -
Arenga nut, - 109 -
ASEAN, - 106 -
Atoin Meto, - 142 -
Atoin Pah Meto, - 142 -
attic, - 39 -, - 108 -, - 109 -
AUSTRONESIAN, - 111 -

B

Badjo, - 139 -
Baduy, - 99 -
Bagas Godang, - 79 -, - 84 -
Baileo, - 128 -
Bajao Laut, - 139 -
Bajaree Chuttong, - 106 -
Balai Agung, - 89 -
balancing, - 89 -
Bali, - 106 -
Balinese, - 82 -, - 106 -, - 125 -
Baloy house, - 120 -
Banda Sea, - 147 -
Bandung, - 58 -
Bandung Institute of Technology, - 58 -
Bangli, - 106 -
Banjar Dayak, - 125 -
Banjarmasin, - 125 -
Banten, - 99 -
Banua, - 117 -
Banua Layuk, - 117 -
Banua Sura, - 117 -
Barabai, - 125 -
Barn roof, - 108 -
Batak Alas, - 75 -
Batak Bebesen, - 75 -, - 76 -
Batak Gayo, - 74 -, - 75 -, - 76 -
Batak Karo, - 2 -, - 72 -
Batak Kluet, - 78 -
Batak Singkil, - 75 -
Batavian, - 94 -
beach huts, - 8 -
bed and breakfast, - 149 -
Bedouin house, - 99 -
Bedui, - 99 -
Bee housing, - 15 -
Beescape, - 15 -, - 109 -
Begonjong, - 88 -
Belltower, - 84 -
Bena, - 137 -
Bengkalis, - 89 -
Bengkulu, - 89 -
Betang, - 121 -
Betawi, - 94 -, - 95 -, - 100 -
Betawi Joglo, - 95 -
Bhaga, - 137 -
Bidayuh, - 122 -
Bima, - 127 -, - 139 -
Bines dance, - 78 -
blind spots, - 8 -
Bloh and Napung Rivers, - 150 -
boat house, - 89 -
Bolon, - 79 -, - 80 -
Borneo, - 119 -
Boti, - 136 -
Boti Village, - 136 -
box, - 66 -
box hives, - 8 -, - 15 -, - 66 -, - 103 -
Boyang House, - 116 -
branch hollow, - 109 -
Brunei, - 16 -, - 37 -, - 40 -, - 45 -, - 64 -, - 67 -, - 69 -
Bubungan Lima, - 89 -
buffalo horns, - 112 -
Buginese, - 118 -
Bugis, - 118 -, - 139 -, - 144 -
bumiputra, - 43 -
Buru Island, - 131 -
Buru people, - 132 -
Buton, - 117 -, - 118 -
Butung, - 118 -

C

cabanas, - 8 -
Cakra Donya, - 83 -
Cancar, - 135 -
canopy, - 83 -
Capit Gunting, - 100 -
carved motifs, - 46 -
Celebes, - 111 -
champions, - 85 -
Chiang Mai, - 59 -
China, - 17 -, - 89 -
Chinese pavilions, - 8 -
Cikgu Aman, - 61 -
Claus Rasmussen, - 8 -
Clerestory, - 109 -
Colonial influence, - 108 -
colony population, - 14 -
concave saddle, - 100 -
Conservation, - 14 -
Coral Triangle, - 139 -
cuboid boxes, - 106 -
curved ridge, - 15 -

D

Dalam Loka, - 126 -
Dawan, - 136 -, - 142 -
Dayak, - 120 -
Dayak house, - 125 -
Deru Lalulewa, - 137 -
Deru Solamae, - 137 -
Dizi Azi, - 137 -
Dompu, - 127 -
Dompu tribe, - 127 -
Doo Island, - 149 -
doorbells, - 39 -
Double pitch, - 15 -
Dragon, - 95 -
Dugong watching, - 150 -
Dulohupa, - 116 -
Dutch, - 108 -

E

Earthquake Resistant, - 143 -
earthquake-resistant, - 89 -
earthquakes, - 109 -
East Nusa Tenggara, - 126 -
East-Flores clan, - 138 -
eduction, - 108 -, - 109 -
Einstein, - 9 -
Ende, - 135 -
Epiphyllum oxypetalum, - 46 -
Ethnic Cultures, - 111 -
evolution, - 57 -
extreme weather, - 14 -

F

feral nests, - 109 -
finial, - 86 -
flood levels, - 39 -
Flores Island, - 134 -, - 137 -
Flores Sea, - 134 -
foraging sources, - 14 -
fruit stalk, - 109 -
fuscobalteata, - 109 -

G

Gable roof, - 15 -
Gajah Merem, - 89 -
gallows, - 137 -
Gambrel, - 147 -
Gambrel roof, - 108 -
Gayo Deret, - 74 -, - 75 -, - 77 -
Gayo Kalul, - 74 -, - 75 -
Gayo Lues, - 74 -, - 75 -, - 76 -
Gayo Lut, - 74 -, - 75 -, - 77 -, - 78 -
Gayo Serbejadi, - 74 -, - 75 -, - 76 -
gazebos, - 8 -
Gebemliar, - 132 -
Gebfuka, - 132 -
Geniotrigona thoracica, - 66 -
Georgian, - 108 -
gongs, - 39 -
Gorontalo City, - 116 -
Gurusina, - 135 -

H

Halmahera, - 129 -, - 131 -
Halmahera Island, - 132 -
Harun, 1991, - 100 -
harvesting, - 66 -
hateup, - 99 -
heat dissipation, - 72 -
Heliana Island, - 149 -
heritage, - 60 -, - 106 -
Heritage Village, - 39 -
Heterotrigona itama, - 61 -, - 66 -
hexagon, - 8 -
Hibualamo, - 129 -, - 130 -
Hindu, - 107 -
Hip, - 126 -
hollowed-out log nest, - 2 -
honey pots, - 66 -
Honey production, - 66 -
Honey Super, - 95 -
honeycomb, - 9 -
honour Allah, - 88 -
Huaulu tribe, - 128 -
Humba, - 141 -

I

ijuk, - 99 -
Ile Ape, - 147 -
Imah adat Sunda, - 99 -
Indian Ocean, - 134 -
indigenous, - 8 -, - 99 -, - 117 -
Indonesia, - 85 -
Inoue, - 85 -
insect tourism, - 8 -
Insect Tourism, - 46 -
interwoven rafters, - 85 -
Islam, - 20 -, - 73 -, - 76 -, - 77 -, - 78 -, - 118 -, - 144 -
Islamic law, - 89 -
Istana Dalam Loka, - 126 -
itama, - 66 -

J

Java, - 57 -, - 94 -, - 95 -, - 96 -, - 99 -, - 100 -, - 103 -
Javanese, - 15 -, - 94 -, - 95 -, - 96 -
Javanese Joglo, - 15 -, - 96 -
Joglo roof, - 97 -
Juha, - 86 -
Julang Ngapak, - 100 -
jungle retreat, - 149 -
Jungle Trek Camp, - 59 -

K

Kaba Ena, - 117 -
Kadato Kie, - 130 -
Kaili, - 117 -
kain, - 149 -
Kajang Lako, - 89 -
Kalamanthana, - 119 -
Kalimantan, - 119 -, - 125 -
Kampar, - 88 -
Kampong Glam, - 62 -
Kampung roof, - 97 -
Kampung style, - 97 -
Karema, - 117 -
Kasepuhan House, - 99 -
Kayeli, - 132 -
Kebaya, - 94 -, - 95 -
Kebayakan, - 76 -
Kelake Ado Pehan, - 145 -
Khopa, - 137 -
Kië ma-kolano, - 131 -
Kinabalu Park, - 39 -
kisanatapat, - 137 -
Klemantan, - 122 -
Komodo National Park, - 139 -
Kon awe, - 117 -
Kota Selatan District, - 116 -
Kupang, - 134 -

L

Laika, - 117 -
Lake Toba, - 72 -
Lamahala, - 144 -
Lamatokan, - 147 -
Lamin houses, - 120 -
Lampung, - 89 -, - 103 -
Lancang, - 88 -
Landu Island, - 149 -
Langkawi, - 57 -
Laut Tawar, - 75 -, - 78 -
Layar and Sayap, - 88 -
Lebak Regency, - 99 -
Lembata, - 145 -, - 146 -, - 147 -
lentik, - 88 -
lesser inclination, - 97 -
Leuit, - 99 -
Lewoleba, - 147 -
Leworahang, - 138 -
Limas, - 88 -, - 89 -
Limba Village, - 116 -
Lisela, - 132 -
Lohayong, - 144 -, - 145 -
Loka, - 126 -
Lokop, - 76 -
Loloan, - 107 -
Loloan Malays, - 107 -
Lombok, - 108 -, - 109 -
Lontiak, - 88 -
Lopo, - 136 -
Lore tribes, - 117 -
Lotud house, - 38 -
Lotud tribe, - 39 -
Lumbung, - 108 -, - 109 -

M

Majo Melayu, - 88 -
Majo Tribe, - 88 -
Makassarese, - 126 -
Malaccan, - 15 -
Malay architecture, - 88 -
Malay Heritage Centre, - 62 -
Malay Peninsula, - 56 -
Malaya Independence, - 94 -
Malays, - 15 -
Malaysia, - 15 -, - 94 -
males, - 66 -
Maluku, - 128 -, - 130 -, - 131 -
Mamasa, - 117 -
Mamuju, - 117 -
Manado, - 114 -
Mandailing, - 79 -, - 84 -
Mandar, - 117 -
Manggaraian houses, - 135 -
Manuk Island, - 149 -
Maranao, - 60 -
Mekongga, - 117 -

Melanau, - 43 -, - 122 -
meliponarium, - 109 -
Meliponary design, - 8 -
Meliponiculture Tourism, - 111 -
menang kerbau, - 85 -
Menanga, - 144 -
MHC, - 62 -
Minahasa, - 113 -, - 116 -
minangkabau, - 85 -
Minangkabau, - 15 -, - 72 -, - 85 -, - 86 -
Mindanao, - 60 -
miniature model, - 15 -, - 94 -
miniatures, - 111 -, - 112 -, - 119 -
minimalist, - 15 -
Misa, - 139 -
Monitor, - 39 -
monitor or clerestory windows, - 39 -
monitor roof, - 2 -
Montasik, - 82 -
motherland, - 94 -
MS hollow sections, - 39 -
Mt. Rinjani, - 109 -
Muna, - 117 -
Murut, - 40 -
Musalaki, - 126 -
Musalaki houses, - 126 -
Museum, - 83 -
Musyawarah, - 89 -
Myanmar, - 38 -

N

Nabawan, - 42 -
Nagekeo, - 145 -
Nasaruddin, - 57 -
natural fibres, - 8 -
Ndao Island, - 149 -
Ngada, - 136 -, - 137 -
Ngadhu pole, - 137 -
Nga'du, - 137 -
Norita W. Pangestika, - 106 -
North Peninsula, - 15 -
Novelty, - 112 -
NuhaNera, - 147 -
Nusa Tenggara, - 125 -, - 126 -, - 134 -, - 141 -, - 142 -, - 143 -, - 144 -
Nusantara, - 15 -, - 57 -
Nuse Island, - 149 -
Nuwo sesat, - 89 -

O

Old House, - 89 -
Oluhuta, - 114 -
Omjol, - 84 -
open cabinet, - 109 -
Orang Asli, - 42 -, - 43 -
Ozie, - 125 -

P

Padang, - 85 -
Pakpak, - 74 -, - 79 -
Palembang, - 89 -
palm leaves, - 82 -, - 89 -, - 99 -
Pamana Island., - 149 -
Pana, - 118 -
Pantar, - 144 -, - 148 -
Pantar Island, - 148 -
Panyabungan, - 84 -
Pavilion, - 89 -
pepung, - 89 -
Peranakan, - 88 -
Peru, - 8 -
Peureulak, - 84 -
Philippines, - 0 -
Piyeung Datu, - 82 -
platform, - 39 -
Polynesian, - 60 -
Polynesian influence, - 60 -
porch, - 95 -
Portuguese, - 16 -, - 17 -, - 44 -, - 113 -, - 116 -, - 126 -, - 135 -, - 144 -, - 145 -
Prehistoric, - 114 -
PROTO MALAY, - 111 -
Pujut District, - 109 -
purwatin, - 89 -
Pusiban, - 89 -
pyramid house, - 89 -

R

Raha, - 117 -
Rajah James Brooke, - 122 -
Ratenggaro, - 141 -
Reje Linge, - 75 -
replica, - 94 -, - 109 -
resin, - 66 -
return path, - 14 -
Riau Province, - 88 -
ridge forms, - 89 -
Rinca, - 144 -
Robbert Dijkgraaf, - 9 -
rock outcrop, - 147 -
Roman Catholic, - 146 -
Rote Island, - 136 -, - 149 -
Rumah Adat, - 84 -, - 126 -
Rumah Ba-anjung, - 119 -
Rumah Banjar, - 119 -
Rumah Boyang, - 116 -
Rumah Bubungan Tinggi, - 119 -
Rumah Gadang, - 86 -
Rumah Joglo, - 100 -
Rumah Kebaya, - 94 -
Rumah Lamin, - 120 -
Rumah Lamo, - 89 -
Rumah Lancang, - 88 -
Rumah Pencalang, - 88 -
Rumah Pewaris, - 116 -
Rumah Tambi, - 117 -
Rumoh Aceh, - 82 -, - 83 -
Rurung Agung, - 90 -
Ruteng, - 135 -

S

Sabah, - 38 -, - 39 -
Sabunese, - 147 -
saddle pattern, - 95 -
saddle roof, - 99 -, - 100 -
sago palm, - 82 -
Sahu language, - 129 -
Sahu tribe, - 129 -
Sakagami, - 85 -, - 86 -
Salutambun, - 118 -
Sama-Bajau, - 140 -
Samosir Island, - 72 -
Sangihe Islands, - 118 -
Sanskrit, - 111 -, - 119 -, - 126 -
Sarawak, - 13 -, - 34 -, - 40 -, - 43 -, - 122 -
Sasadu, - 129 -
Sasak, - 108 -, - 109 -
scent, - 14 -
Science Festival, - 9 -
scissors, - 99 -, - 100 -
sea gipsies, - 139 -
Sebuku, - 78 -
Selembayung, - 88 -
Semau, - 144 -
Semawa, - 126 -
Senna sumatrana, - 86 -
Sika, - 150 -
Sika people, - 150 -
Simalungun, - 79 -
Singapore, - 16 -, - 62 -, - 104 -, - 125 -
Siti Salmah, - 85 -
Sofifi, - 131 -
Solor Islands, - 143 -, - 144 -
spire, - 86 -, - 100 -
Stilt house, - 95 -
stilts, - 39 -, - 99 -, - 109 -, - 116 -
stone pedestals, - 39 -
store grain, - 108 -, - 109 -
straw hat, - 149 -
strong gales, - 14 -
sturdy pole, - 86 -
Sulah Nyanda, - 99 -
Sulawesi, - 111 -, - 116 -, - 117 -
Sultan Muhammad Jalalludin III, - 126 -
Sumatera, - 85 -
Sumatra, - 72 -, - 82 -, - 84 -, - 85 -, - 86 -, - 88 -, - 89 -, - 103 -
Sumba, - 134 -, - 141 -, - 143 -

Sumba Island, - 141 -
Sumbawa, - 125 -, - 126 -, - 127 -, - 141 -
Sumbawa palace, - 126 -
Sunda, - 99 -, - 100 -
Sunda Islands, - 2 -, - 130 -, - 134 -, - 143 -
Sundanese, - 99 -, - 100 -
symmetry, - 9 -

T

Tagog Anjing, - 100 -
Takpala, - 148 -
Tana Toraja Regency, - 111 -
Tanduk Buang, - 88 -
Tanjung District, - 109 -
Tasik Merimbun, - 66 -
tathu, - 132 -
terminata, - 66 -
Ternate, - 118 -, - 129 -, - 131 -
Terong, - 144 -
Terrarium, - 109 -
Tetabuhan, - 89 -
Tetragonula, - 66 -, - 85 -
Tetragonula drescheri, - 103 -
Tetragonula fuscobalteata, - 66 -, - 108 -
Tetragonula minangkabau, - 85 -
Tetrigona binghami, - 66 -
Thailand, - 16 -, - 44 -, - 59 -

thermal comfort, - 13 -, - 100 -, - 112 -
Tidore, - 130 -, - 131 -
Tidung, - 120 -
Timor, - 16 -, - 130 -, - 134 -, - 136 -, - 142 -, - 143 -, - 144 -
Timor Island., - 136 -, - 150 -
Timorese Sika, - 150 -
Toba Batak, - 72 -
Tobelo people, - 132 -
Todo, - 135 -
Tolaki tribe, - 117 -
Tolitihu, - 116 -
Tolotang, - 118 -
Tonga, - 84 -
Tongkonan, - 111 -
Top Hats, - 142 -
Toraja, - 111 -, - 112 -
Torajans, - 111 -
Totem, - 125 -
transition store, - 109 -
Tree House, - 39 -
Tuaran, - 39 -
tunjuk langit, - 86 -

U

Uab Meto, - 142 -
Ulu Tutong, - 66 -
Uma ceko, - 127 -
Uma Jompa, - 127 -
Uma Kelada, - 141 -
Uma Lengge, - 127 -
Uma Mbolo, - 127 -
Uma Panggu, - 127 -
ume kbubu, - 136 -, - 142 -

V

ventilation, - 13 -, - 18 -, - 19 -, - 21 -, - 44 -, - 102 -
VOC, - 144 -, - 145 -

W

Wae Rebo, - 135 -
Wahto, - 137 -
Wai Apu people, - 131 -
Waka Hamlet, - 117 -
Walewangko, - 116 -
Wallacea, - 108 -
Walu Mandoku, - 141 -
West Nusa Tenggara, - 126 -
Wetar, - 130 -
Wim Sutrisno, - 83 -
Wolio, - 117 -
Wologai, - 135 -
Wolotopo, - 135 -
Wolowaru, - 135 -
wooden houses, - 15 -
woven bamboo, - 82 -, - 99 -

Bibliography

Adnan, W. N., Sajap, A. S., Adam, N. A., & Hamid, a. M. (2015). Flight Intensity of Two Species of Stingless Bees *Heterotrigona itama* and *Geniotrigona thoracica* and Its Relationships with Temperature, Light Intensity and Relative Humidity. *Centre for Insects Systematic, UKM.*

Anggraeni, D. (2011). Another East: Representation of Papua in Popular Media. *Conference: International Conference on Indonesian Studies at the University of Indonesia, Depok, Indonesia.*

Appanah(eds), S., & Turnbull(eds), J. M. (1998). A Review of Dipterocarps: Taxonomy, ecology and silviculture. *FRIM.*

Bahauddin A. & Musadat D. M. S. A,. (2018). The Traditional Architecture of the Melanau Tall Longhouse, Mukah, Sarawak. S*HS Web of Conferences* 45, 01002.

Barfod, A. S., Hagen, M., & Borchsenius, F. (2011). Twenty-five years of progress in understanding pollination mechanisms in palms (Arecaceae). *Annals of Botany, Volume* 108, *Issue* 8.

Birt., M. (n.d.). The Trigona Species, (Stingless Bees). Their role in higher plant pollination. *12391833933.*

CORTOPASSI-LAURINO, M., & al, e. (2006). Global meliponiculture: challenges and opportunities. *Apidologie 37*, 275–292.

Cossio, B. M. (2005). The Buka–Hatene Community Learning Centre: Friends of Baucau's Project to Rebuild a Community Building in Baucau, Timor Leste. The *University of Chile.*

Couvillon, M. J rt al. (2008). A comparative study in stingless bees (Meliponini) demonstrates that nest entrance size predicts traffic and defensivity. *J . EVOL. BIOL. 21*, 194–201.

Danaraddi, C. S., Viraktamath, S., & Bhat, K. B. (2009). Nesting habits and nest structure of stingless bee, Trigona iridipennis Smith at Dharwad, Karnataka. *Karnataka J. Agric. Sci., 22(2):,* (310-313) .

Darmawan, S. (n.d.). Diversifikasi Produk *Trigona* sp. Di Lombok. *Badan Litbang Kehutanan, Kementrian Kehutanan.*

Eltz, T., & al., e. (2003). Nesting and Nest trees of stingless bees (Apidae; Meliponini) in lowland dipterocarp forest in Sabah, Malaysia with implications for forest management. *Forest Ecology and Management 172.*

Faisal, G. (2019). Arsitektur Melayu: Rumah Melayu Lontiak Suku Majo Kampar. *Jurnal Arsitektur, Vol. 6, No. 1.*

Fatoni, A. (2008). Pengaruh propolis *Trigona* spp. asal Bukittinggi terhadap beberapa bakteri usus halus sapi dan penelusuran komponen aktifnya [Tesis]. Bogor (ID):. *Program Pascasarjana, Institut Pertanian Bogor, Bogor.*

Hadi S., Ziegler T. & Hodges K. . (2009). Group Structure and Physical Characteristics of Simakobu Monkeys (Simias concolor) on the Mentawai Island of Siberut, Indonesia. *PubMed.*

Hanifa H S et al. (2021). Characteristics of apiculture and meliponiculture in Banten Province, Indonesia: profile of beekeepers bee and pollen diversity. *IOP Conf. Ser.: Earth Environ. Sci.* 948 012050.

Hong, N. (2018). Tanimbarese and Rotinese Traditional House. The *National University of Singapore.*

Hussain, N. H. (2015). Thoughts on Malaysian Architecture Identity and Design Principles of Malayan Architects Co-Partnership. *Universiti Teknologi Malaysia.*

Iyengar, K. (2015). Sustainable Architectural Design. *Routledge.*

Jalil, A. (2014). Beescape for Meliponines. *Singapore: Partridge Publishing.*

Jalil, A. H. (2019). Cohabitation of Meliponines in Ants' or Termites' Arvoreal nests. *Alademi Kelulut Malaysia Sdn Bhd.*

Jalil, A., & Roubik(ed), D. (2016). Handbook of Meliponiculture. *Akademi Kelulut Malaysia.*

Jalil, A., & Roubik(ed), D. (2018). Handbook of Meliponiculture Vol 2. *Akademi Kelulut malaysia.*

Jalil, A., & Roubik(ed), D. (2021). MALAYSIAN MELIPONICULTURE & BEYOND Inc. Stingless Bee Conservation. *UK: IBRA & NBB.*

Jaltl A. H. & Roubik(Ed) D. W. (2022). Indonesian Meliponiculture & Beyond. *UK: IBRA & NBB.*

Janra M. N., Herwina H., Salmah S., Rusdimansyah & Jasmi. (2020). Identification of Potential Predators and Pests in Stingless Bee Farm (Hymenoptera; Apidae; Meliponini; Tetragonula, Lepidotrigona) through Rapid Observation in Padang Pariaman Regency, West Sumatra. *Jurnal Sumberdaya HAYATI Vol. 6 No. 2, November*, 67-74.

Lamerkabel, J. S., & et al. (2017). Tabiat Bersarang Dua Spesis Stingless Bee Koloni Di Pulau Ambon Muluku. *Seminar Nasional Perlebahan.* Bogor: IPB.

Leonhardt, S. D., & al., e. (2010). Stingless Bees Use Terpenes as Olfactory Cues to Find Resin Sources. *Chem. Senses 35*, 603–611.

Michener, C. D. (1961). Observations on the Nests and Behavior of Trigona in Australia and New Guinea (Hymenoptera, Apidae). *AMERICAN MUSEUM NOVITATES.*

Nanuru R. F., Munir M. & Tjahyadi S. . (2019). Sasadu: The Religious Social Spirit of Sahu Tribe Community in North Maluku - Indonesia. *The Journal of Social Sciences Research Vol. 5, Issue. 4*, 1274-1283.

Nas P. J. M., Shahab Y. Z. & Wuisman J. J. J. M. . (2008). The Betawi house in Jakarta. In T. N. Leiden, *Indonesian Houses Vol. 2* (pp. 597–628). BRILL.

Nijel H. (2018). Tanimbarese And Rotinese Traditional House. The *National University of Singapore.*

Ntawuzumunsi E., Kumaran S. & Sibomana L. . (2021). Self-Powered Smart Beehive Monitoring and Control System (SBMaCS) †. *Sensors, 21*, 3522.

Palittin D. & Hallatu T.G.R. (2019). *Sar: Kanume tribal culture in environmental conservation to reduce global warming effects.* IOP Conf. Ser.: Earth Environ. Sci. 235 012062.

Phoek I. C. A.; Tjilen A. P. & Cahyono E. . (2021). *Analysis of Ecotourism, Culture and Local Community Empowerment: Wasur National Park - Indonesia Case Study.* Macro Management & Public Policies | Volume 03 | Issue 02.

Putra, B. A. (2019). Cultural Representation of Vernacular Housing in Melayu Jambi Traditions. *Journal V-Tech (Vision Technology Vol.2 NO. 1.*

Rahim M., Ibrahim M. & Marasabessy F. . (2021). Construction System and Environment Adaptation of Traditional Architecture in Moluccas Island. *Civil Engineering and Architecture 9(5):*, 1530-1545.

Rasmussen. (2008). Catalogue of the Indo-Malayan/Australasian stingless bees (Hymenoptera: Apidae: Meliponini).

Rasmussen, C. (2013). Stingless bees (Hymenoptera: Apidae: Meliponini) of the Indian subcontinent: Diversity, taxonomy and current status of knowledge. *Zootaxa 3647 (3): 401–428.*

Riendriasari, S. D. (2014). Budidaya dan Produk Perlebahan *Trigona* Spp Di Lombok, Nusa Tenggara Barat. *Prosiding Seminar Nasional* (pp. 213 -221). *Jogja: University Club Universitas Gadjah Mada,.*

Roubik, D. (1995). Pollination of cultivated plants in the tropice. *FAO Agric. Serv. Bull. 118, Rome.*

Roubik, D. (2006). Stingless bee nesting biology. *Apidologie 37.*

Sakagami, S. F. (1978). Tetragonula Stingless Bees of Continental Asia and Sri Lanka (Hymenoptera, Apidae). *Jour. Fac. Sci. Hokkaido Univ. Ser. VZ, 2001. 21(2).*

Sakagami, S. F., Inoue, T., Yamane, S., & Salmah, S. (1983). Nest architecture and colony composition of the Sumatran stingless bee *Trigona (Tetragona) laeviceps. Kontyu, 51(1):*, 100-111.

Salatnaya, H. (2019). Potential growth of meliponiculture in West Halmahera*, Indonesia.* IOP Conference *Series Earth and Environmental Science* · December 2019.

Sayustia, T., Raffiuddin, R., Kahono, S., & Nagir, T. (2021). Stingless bees (Hymenoptera: Apidae) in South and West Sulawesi, Indonesia: morphology, nest structure, and molecular characteristics. *Journal of Apicultural Research Volume 60, 2021 - Issue 1.*

Schwarz, H. F. (1937). Bornean Stingless Bees of The Genus *Trigona. Bulletin American Museum of Natural History.*

Schwarz, H. F. (1939). The Indo Malayan Species of Trigona. *Bulletin of AMNH.*

Sim, S. (2010). Redefining the Vernacular in the Hybrid Architecture of Malaysia. *Victoria University of Wellington.*

Sommeijer, M. (1999). Beekeeping with stingless bees: a new type of hive. *Bee World 80(2):* 70-79.

Syafrizal, F., Tarigan, D., & Yusuf, R. (2014). Biodiversity and Habitat of Trigona at Secondary Tropical Rain Forest of Lempake Education Forest, Samarinda, Kalimantan Timur. *Jurnal Teknologi Pertanian 9(1):34-38,.*

Tjoa-Bonatz M. L., Neidel J. D., & Widiatmoko A. . (2009). Early Architectural Images from Muara Jambi on Sumatra, Indonesia. *Muara Jambi: Asian Perspectives*, Vol. 48, No. 1.

Vit, P., Roubik, D. W., & Pedro, S. M. (2013). Pot Honey - A Legacy of Stingless Bees. *Springer*, DO - 10.1007/978-1-4614-4960-7.

Vit, P.; Pedro, S.R.M.; Roubik, D. (2018). Pot-Pollen in Stingless Bee *Melittology. Springer International Publishing AG.* DO - 10.1007/978-3-319-61839-5.

Welzen, P. C., J. F., & Alahuhta, J. (2005). Plant distribution patterns and plate tectonics in Malesia. *Biologiske Skrifter 55*, 199-217.

Welzen, P. C., Parnell, J. A., & Slik, F. (2011). Wallace's Line and plant distributions: Two or three phytogeographical areas and where to group Java? *Biological Journal of the Linnean Society*.

Wikipedia. (5 Desember 2018). *Daftar_kayu_di_Indonesia*. Retrieved from id.wikipedia.org: https://id.wikipedia.org/wiki/Daftar_kayu_di_Indonesia

Xiong, C. S. (2015). A Conservation Assessment of Stingless Bees (Apidae: Meliponini) in Singapore. *Cohort AY2014/2015 S1*.

Zain, Z. (2012). Analisis Bentuk dan Ruang Pada Rumah Melayu Tradisional Di Kota Sambas Kalimantan Barat. *NALARs Volume 11 Nomor 1*, 39-62.

www.ingramcontent.com/pod-product-compliance
Lightning Source LLC
Chambersburg PA
CBHW061551010526
44117CB00022B/2987